OPTICAL WAVEGUIDE MODES

ABOUT THE AUTHORS

Richard J. Black, Ph.D., is a leading authority on optical waveguide modes and applications to optical fiber sensing problems and structural health monitoring. He has more than 25 years of experience in photonics in both academia and industry, and has coauthored more than 250 publications and official technical reports. Dr. Black is a founding member and Chief Scientist of Intelligent Fiber Optic Systems Corporation (www.ifos.com) and founder of OptoSapiens Design (www.optosapiens.com). He is a lifetime member of OSA and ASM International, a senior member of IEEE, and a member of SPIE, AAAI, ACM, and SAMPE.

Langis Gagnon, Ph.D., is Lead Researcher and Team Director for Vision and Imaging at CRIM (Centre de Recherche Informatique de Montréal) and Adjunct Professor of Computer and Electrical Engineering at Université Laval. He has published more than 150 scientific articles relating to the fields of optics, image processing, object recognition, and math-based nonlinear optical modeling, including solitons and symmetry groups of differential equations. Dr. Gagnon is a member of SPIE, ACM, IEEE, AIA, and IASTED.

OPTICAL WAVEGUIDE MODES

Polarization, Coupling, and Symmetry

RICHARD J. BLACK, PH.D.

OptoSapiens Design

LANGIS GAGNON, PH.D.

Centre de Recherche Informatique de Montréal

McGraw Hill

New York Chicago San Francisco Lisbon London Madrid
Mexico City Milan New Delhi San Juan Seoul
Singapore Sydney Toronto

The McGraw-Hill Companies

Cataloging-in-Publication Data is on file with the Library of Congress

1 2 3 4 5 6 7 8 9 0 WFR/WFR 1 6 5 4 3 2 1 0

ISBN 978-0-07-162296-7
MHID 0-07-162296-9

Sponsoring Editor
Taisuke Soda

Editing Supervisor
Stephen M. Smith

Production Supervisor
Pamela A. Pelton

Acquisitions Coordinator
Michael Mulcahy

Project Manager
Shruti Vasishta, Glyph International

Copy Editor
Patti Scott

Proofreader
Lucia Read

Art Director, Cover
Jeff Weeks

Composition
Glyph International

Printed and bound by Worldcolor/Fairfield.

This book is printed on acid-free paper.

CONTENTS

Chapter 3

Circular Isotropic Longitudinally Invariant Fibers 35

Chapter 4

Azimuthal Symmetry Breaking 67

Chapter 5

Birefringence: Linear, Radial, and Circular 83

PREFACE

This book is about the modes of single- and few-mode optical waveguides with an emphasis on single-core and multicore optical fibers and couplers including a large range of geometries and anisotropies, both standard and exotic. It provides both an "atlas" of modal field forms and an understanding of the physical properties resulting from waveguide symmetries. In addition to optical waveguide and fiber-optic designers, researchers, and students, this book may appeal to quantum and solid-state chemists and physicists interested in the application by analogy of techniques they know well in the continually expanding field of photonics.

To aid in rapid understanding, we emphasize a building-block approach with approximate modes and simplified structures forming a basis for more exact analyses and more complex structures. Accordingly we commence with single-core fibers and the symmetry consequences arising from specific forms of the azimuthal and radial dependence of the index profile.

The mathematical tools involve (1) the weak-guidance perturbation formalism facilitating the incorporation of polarization effects following a scalar analysis together with (2) a group theoretic approach for systematic exploitation of symmetry.

Scalar modes provide a basis for vector modes. Field constructions for transverse and hybrid polarized modes in terms of both linearly and circularly polarized modes are given. Degeneracy splittings and vector mode field transformations are considered depending on the relative strengths of the refractive index profile height and deformations from a circular cross section (e.g., elliptical, triangular, square) or birefringence (linear, radial or azimuthal, circular). Both microscopic and macroscopic anisotropies are considered: The polarization effects arising from a single interface may be regarded as a macroscopic manifestation of form birefringence. Single-core results are then used as a building block in the analysis arrays of few-mode lightguides: multicore fibers and multifiber couplers.

The organization of material is as follows:

- Chapter 1 provides an introduction including a motivation for the study of waveguide mode forms.
- Chapter 2 starts from the fundamental Maxwell equations for electrically anisotropic and isotropic media to provide a

comprehensive treatment of the resulting wave equations. For longitudinally invariant optical waveguides, it emphasizes the weak-guidance formalism which in general leads to perturbation expansion in terms of the typically small fractional refractive index difference between the waveguide core and cladding.

- Chapter 3 considers the scalar and vector modes of circular optical fibers. It includes a tutorial introduction to the consequences of symmetry, using a group theoretic approach in degeneracy determination and field construction of different modes of circularly symmetric fibers.
- Chapter 4 examines elliptical, triangular, and square deformations of circular waveguide cross sections as illustrations of the modal degeneracy splitting and field transformation resulting when the *azimuthal* circular symmetry is lowered to n-fold rotation-reflection symmetry.
- Chapter 5 considers linearly, radially, and circularly birefringent (gyrotropic) fibers.
- Chapter 6 is devoted to the construction of modes of multicore fibers and multifiber coupler arrays.
- Chapter 7 provides a summary of the results and discusses extensions of the concept of modes for longitudinally invariant structures to modes for structures with longitudinal variations, such as periodic structures and Kerr-type nonlinear waveguides where intensity-dependence induces longitudinal variation in the presence of a propagating wave.
- The appendix provides the essential results of elementary applied group representation theory used for the analysis of many physical and chemical systems involving symmetry. Together with the symmetry tutorial included in Sec. 3.2, this provides an alternative introduction to and/or illustration of concepts which students might apply by analogy in many other fields such as quantum, solid-state, and molecular chemistry and physics.

Richard J. Black
Langis Gagnon

ACKNOWLEDGMENTS

Following some inspiring discussions and correspondence with Prof. Geoff Stedman in 1986, this book had its origins in two manuscripts [1, 2] prepared by us in the mid to late 1980s while Richard J. Black (RJB) was at the École Polytechnique de Montréal and Langis Gagnon (LG) was at the Université de Montréal. Following the encouragement of Prof. Carlo Someda, the first full version of this book was prepared in 1991–1992, with relevant references up until that time, while RJB was at the École Polytechnique de Montréal. It was later revised for part of a course presented by RJB at the Swiss Federal Institute of Technology [École Polytechnique de Lausanne (EPFL)] in January–February 1995. The present 2010 revised version followed from discussions between RJB and Taisuke Soda of McGraw-Hill, who we thank, together with all the McGraw-Hill and Glyph International team, particularly Shruti Vasishta, for expert preparation of the book. We also thank colleagues at Photon Design, Technix by CBS, IFOS, and CRIM, and many other colleagues, family, and friends too numerous to mention, for their contributions and support.

The present new version includes (a) a simple intuitive introduction to waveguide modes (Sec. 1.5) aimed at those encountering them for the first time, (b) recent developments (Secs. 7.4 through 7.8), and (c) 78 additional references. (With regard to references, the first 138 appeared in the 1992 manuscript, and Refs. 139 and 140 were added for the 1995 manuscript.) While the fundamental theory of optical waveguide modes presented herein remains the same, since 1992 we have witnessed considerable growth in photonics in the commercial sector (particularly rapid in telecom in the late 1990s with steady progress in photonic sensors to the present), with technical and scientific developments in many areas, for example, periodic lightguides [fiber Bragg gratings (FBGs), photonic crystal fibers, and photonic crystals] and waveguide modeling packages. We touch on these areas in added Refs. 141 and above together with new Secs. 7.4 through 7.8.

Richard J. Black
Langis Gagnon
2010

RJB is grateful to Prof. René-Paul Salathé and the Swiss Federal Institute of Technology, Lausanne (EPFL), for providing him with the opportunity to present this material as a course at EPFL, and to those who suggested improvements to the 1992 version, particularly Prof. Carlo Someda.

Richard J. Black
1995

We are especially indebted to Prof. Geoffrey E. Stedman, University of Canterbury, New Zealand, for many very perceptive, enlightening discussions and correspondence that provided much of the initial insight and inspiration. We thank Dion Astwood for undertaking a useful student project [94] to clarify points regarding modal transformation properties and splittings. Many others kindly provided ideas, discussions, and support. We are particularly grateful to Prof. Carlo Someda for making this book possible with his very generous contribution of kind and patient correspondence and many ideas that helped to improve the original manuscript as well as the much appreciated hospitality to RJB at the Università di Padova; Prof. George Stegeman for hospitality to RJB at the Optical Sciences Center, University of Arizona, during the initial stages; and Prof. John Shaw of Stanford for extensive hospitality and many discussions regarding few-mode fiber devices. We thank Dr. Ken Hill, Communications Research Center (CRC), Canada, and Dr. Richard Lowe and Costas Saravanos of Northern Telecom for support and discussions regarding modes in couplers, tapered fibers, and modal interferometry; and Dr. Iain M. Skinner, University of New South Wales (formerly of CRC), for comments and enlightening discussions regarding mode transitions [20]. RJB thanks his colleagues at the École Polytechnique de Montréal, Profs. Jacques Bures and Suzanne Lacroix, and Dr. François Gonthier for their support and ideas regarding multifiber couplers and modal interferometry, and Profs. John Harnad and Pavel Winternitz for hospitality at the Centre de Recherches Mathématiques (CRM), Université de Montréal, and discussions regarding nonlinear fibers and group theory, as well as the Australian waveguide theorists Profs. Colin Pask, Allan Snyder, and John Love during RJB's formation of ideas on waveguide modes. LG thanks Prof. Pavel Winternitz, Université de Montréal, and Prof. Pierre-André Bélanger, Centre d'Optique Photonique et Laser, Université Laval, for helpful discussions and support.

Richard J. Black
Langis Gagnon
1992

OPTICAL WAVEGUIDE MODES

CHAPTER 1

Introduction

In this chapter, Secs. 1.1 through 1.4 introduce the major themes of the book. Section 1.5 provides an intuitive introduction to optical waveguide modes and Sec. 1.6 provides a chapter-by-chapter outline of the remainder of the book highlighting major results.

1.1 MODES

This is a book about *lightguide mode forms*. In particular,

1. We emphasize the basic structure of **modal field patterns** *in optical fiber cross sections* **transverse** to the direction of propagation.
2. We consider the relative **longitudinal** dependencies of modal fields in terms of their *propagation constant degeneracies or splittings*. Our major objective is to provide an understanding of **how transverse optical waveguide geometry influences modal polarization properties** with refractive index variations ranging from macroscopic "global anisotropies" down to scales much smaller than a wavelength where the local refractive indices of the constituent waveguide media can be treated as anisotropic. As in Refs. 1 and 2, we undertake the analysis using extensions of the weak-guidance perturbation formalism [3] together with elementary group representation theory [4–6]; see also Refs. 7 through 10.

As well as providing the basic general electromagnetic formalism and structural description appropriate for analysis of the lowest-order or fundamental modes (i.e., the two polarization states of the modes referred to as HE_{11} and LP_{01} or CP_{01}), we go beyond that mainstay of present-day long-distance telecommunications and include a detailed introduction and classification of diverse forms of higher-order modes and various polarization manifestations thereof, e.g., modes of polarization that are transverse electric (TE), transverse magnetic (TM), hybrid (HE or EH), linear (LP), circular (CP), and "triangular" (TP). We give particular attention to the *second-order modes*; e.g., for circular fibers, these are the TE_{01}, TM_{01}, and HE_{21} modes, each of which may be constructed in terms of two linearly polarized (LP_{11}) "pseudo-modes" or alternatively in terms of circularly polarized (CP_{11}) modes.

Apart from the applications, since the original circular fiber modal classification scheme due to Snitzer [11], few-mode lightguide problems have attained a particular physical interest in their own right, e.g., Ref. 12. Indeed, our major aim is simply **to provide an understanding of the fundamental physics of mode structure**. It is our belief that a valuable basis for future novel waveguide designs and exploitations will be provided by a thorough knowledge of how waveguide structure—ranging from standard to exotic—can be used to create and manipulate modes with the desired properties.

While *we mostly restrict ourselves to the concept of monochromatic independently propagating modes of idealized lightguides with longitudinally invariant linear refractive indices*, **these ideal "linear" modes** may form the **basis** for **adaptations to perturbed and other less idealized situations** including **longitudinal variations** and **"nonlinear" effects** using coupled-mode, local-mode, coupled-local-mode, and other approaches [e.g., 3, 13]. In the context of few-mode fibers, we mention but some of the adaptations of topical interest for which a full understanding of ideal linear guide modes as a fundamental building block can provide useful added insight:

1. Nonlinear (Kerr-type) intensity-dependent modal interferometry [14–17]
2. Nonlinear (second harmonic) frequency conversion via phase matchings of different-order modes [18]
3. (Permanent-) Grating induced frequency conversion and filtering [19, 20]

In general, few-mode lightguides have received attention ranging from visual photoreceptor studies [21] to a particular recent interest in **modal interferometry** [22, 23] and applications thereof, such as

1. *Fiber characterization* [24, 25]
2. Few-mode fibers for *sensor application* [26–28]
3. Various *special optical fibers* and other exotic waveguides fabricated with optical materials or metal-optic combinations [12]
4. *Modal interferometric waveguide components* [29] such as
   - Tapered fiber devices [30–32]
   - Filters [33–35]
   - Frequency shifters [36]
   - Mode converters [19]
   - Few-mode fiber interferometric switches [e.g., 37, 38]
   - *Couplers consisting of two or multiple cores or fibers* [39–42] in passive or active form. We regard the majority of couplers as two- or few-mode interferometric devices in that the power exchange between the individual cores (or fibers) results from a beating of the "supermodes," i.e., the normal modes of the total lightguide structure, which, in a first approximation, may be constructed as simple combinations of the modes of the individual cores. A second class of coupler (noninterferometric) is based on the idea of manipulation of local supermode form along the guide. Again, an understanding of the influence of transverse waveguide geometry on mode form can provide a valuable conceptual tool for innovative design.

1.2 POLARIZATION DEPENDENCE OF WAVE PROPAGATION

In addition to **birefringence due to electrical anisotropy** (in crystals), or **stress-related birefringence, any variation of refractive index** may lead to **polarization dependence of wave propagation.**

"Microscopic" index variation that is oscillatory and rapid on the wavelength scale provides the well-known **form birefringent** manifestation of **anisotropy** as in the model of multiple thin parallel plates [43 (Sec. 14.5.2)]. In the classical analysis, a local plane wave incident on the plates (i.e., a medium with refractive index alternating in the plane transverse to the direction of propagation) travels with different phase velocities and thus "sees" different average or effective refractive indices, depending on whether the electric field orientation is perpendicular or parallel to the plates.

"Macroscopic" index variations can also provide **effective anisotropy**. Consider a waveguide in which each mode propagates with a particular phase velocity v_φ (associated with a propagation constant β) and thus sees an associated modal effective refractive index $n_{eff} = c/v_\varphi$, which may depend on polarization.

1. The simplest example is that of a planar or slab guide, i.e., a single parallel plate, but now with dimensions on the order of or greater than the wavelength. There is a splitting between the effective indices seen by modes with electric field polarization parallel (TE_m) and those with perpendicular (TM_m) polarization.
2. Going to two-dimensional cross sections, *geometrical anisotropy* occurs if, e.g., the waveguide is elliptical. Starting with the two polarizations of the fundamental (HE_{11}) mode, effective indices now depend on whether polarization is aligned with the major or minor axis. Thus such structures considered globally are optically anisotropic, even though the constituent core and cladding indices are separately isotropic.
3. Even a perfectly circularly symmetric fiber exhibits a degree of birefringence. For example, while the fundamental HE_{11} modes are now polarization degenerate, there is a splitting between the effective indices of modes that are radially polarized (TM_{0m}) and azimuthally polarized (TE_{0m}). In all cases, the splitting magnitude depends on the core-cladding refractive index difference [3 (Chaps. 12 and 14)] as well as guide dimensions with respect to wavelength. The latter case provides the macroscopic single-interface limit for azimuthally anisotropic media created by a series of concentric rings of alternating index [44, 45], i.e., the cylindrical analog of multiple parallel plates which can be used to augment the azimuthal/radial (TE_{0m}/TM_{0m}) polarization-dependent splitting and possibly eliminate even the hybrid polarized HE_{11} mode.

1.3 WEAK-GUIDANCE APPROACH TO VECTOR MODES

For small index variations or small interface fields, considerable simplification is achieved by considering the scalar wave equation rather than the full Maxwell equations. The resulting scalar modes provide the amplitude of Gloge's linearly polarized (LP) pseudo-modes [46]. Snyder and coworkers [3, 47, 48] pioneered a perturbation approach parameterized in terms of the usually small refractive index profile height and showed how to construct the true vector modes of circular, elliptical, and anisotropic [49] optical waveguides in terms of linearly polarized components. In particular, while elliptical and anisotropic fiber modes are approximately LP, on circular isotropic fibers, only modes of zeroth-order azimuthal symmetry are LP (to lowest order); higher-order modes thereon are constructed from two LP components. Herein we include a formal basis for their intuitive and heuristic constructions for few-mode fibers. These constructions provide a basis for expressions they developed for modal eigenvalue corrections and thus polarization-dependent phase velocity splittings, which in turn lead to field corrections showing, e.g., the degree of field line curvature for those modes which are LP to lowest order. Sammut et al. [50] and Love et al. [51] have provided extensive treatments of the effect of profile grading and ellipticity on the fundamental mode polarization corrections to modal eigenvalues and fields. Here we include particular consideration of few-mode fiber polarization mode splitting [52 (Chap. 4)].

1.4 GROUP THEORY FOR WAVEGUIDES

Group theory provides a well-accepted tool for systematic exploitation of symmetries with applications particularly in applied mathematics (solution of nonlinear differential equations [53]—see also Ref. 54 and references therein), chemistry [55], and physics [e.g., 6–9, 56, 57] including classical, relativistic, quantum [58], particle, field, and solid-state [7, 56, 57] physics as well as nonlinear optics (Refs. 54 and 56 and references therein). For example, especially in solid-state and quantum physics, extensive use is made of *group character* and *representation* theory in the analysis of atomic, molecular, and crystal states: determination of energy level degeneracies and splittings, wave functions, selection rules, etc. These are the methods that we use to exploit symmetries arising in modal analysis of optical waveguides [1, 2].

Group theoretic considerations for waveguides were initially employed in *microwave theory*: analysis of symmetric metallic waveguide junctions starting in the 1940s [59 (Chap. 12), 60, 61] and with subsequent exploitation regarding mode forms, particularly by McIsaac and coworkers [62, 63], for metal-clad waveguides of complex cross-sectional shapes [64, 65] and periodic metallic waveguides [66] including treatments in the 1980s by Preiswerk et al. [67, 68] of DFB lasers of helical and other periodicity.

With regard to *optical waveguides*, group theoretic ideas were initially used for the construction of *single-core* fiber modes by Tjaden [69], and they may be exploited to provide a rigorous derivation of the various vector mode types of single-core fibers [1], as in this chapter where we emphasize symmetry reductions. Such a symmetry viewpoint provides an interesting introduction to optical fibers for those more familiar with quantum-mechanical problems as well as an alternative understanding for the fiber theorist. Novel single-core fiber theory results obtained using this approach [1] include the weak-guidance construction of the circularly polarized modes that were introduced by Kapany and Burke [70], a systematic treatment of modal field evolution given competition between several perturbations, as well as an understanding and classification of radially/azimuthally anisotropic mode forms.

With regard to *multicore* fibers, group theoretic tools were shown to provide a powerful and simplifying approach in a pioneering study by Yamashita and coworkers [71–73]. Primary interest was in weakly coupled single-mode cores in arrays with discrete rotation-reflection ($\mathbf{C}_{nv}$) symmetry (with and without a central core): the vector supermodes (which essentially correspond to combinations of the fundamental HE_{11} modes of the individual step profile cores) were numerically determined as a series expansion using basis functions generated by application of a group *projection operator* method to the *longitudinal* field component. In a weak-guidance framework, as exploited in this chapter and discussed in detail in Ref. 12, a combination of the projection operator generation and the symmetry reduction methods introduced for single-core fibers is particularly profitable for didactic purposes and appropriate for adaptation of the analysis to arrays of *few-mode* cores for which a competition between parameters can lead to different mode forms. Application of the projection operator may be (1) to the individual fiber scalar modes to generate scalar supermodes followed by weak-guidance symmetry reduction construction of the vector supermodes or (2) directly to the *transverse* field components to obtain the vector supermodes. Fields and propagation constant degeneracies may be compared with those for a single-core with symmetry reduced by a $\mathbf{C}_{nv}$ perturbation. This approach also leads to a novel analysis and classification of the modes of multicore fibers with discrete global radial/azimuthal anisotropy.

1.5 OPTICAL WAVEGUIDE MODES: A SIMPLE INTRODUCTION

In this subsection, we provide a simple intuitive introduction to optical waveguide modes. This is in contrast to Chaps. 2 and 3 where we provide a more mathematical description starting from Maxwell's equations for generalized anisotropic media.

1.5.1 Ray Optics Description

The simplest examples of waveguides occur when (1) a planar "core" medium of higher refractive index is sandwiched between two media of lower refractive index, or (2) a fiber core is embedded in a cladding of lower refractive index. As illustrated schematically in Fig. 1.1 for a step index waveguide, if the local propagation direction, θ_z, of a ray is close to that of the waveguide axis (or more precisely, if θ_z is less than the complementary critical angle, θ_c), then by total internal reflection, for an ideal guide, bound light rays bounce along the waveguide without interface loss.

FIGURE 1.1

Guidance of bound and refracting rays along a simple optical waveguide whose refractive index profile is "step index" as shown on the right: (*a*) A ray is bound if the ray propagation angle θ_z is less than the complementary critical angle $\theta_c \approx [1 - (n_{cl}/n_{co})^2]^{1/2}$. (*b*) Refracting rays occur for angles θ_z greater than θ_c. For these rays loss occurs at each bounce along the waveguide.

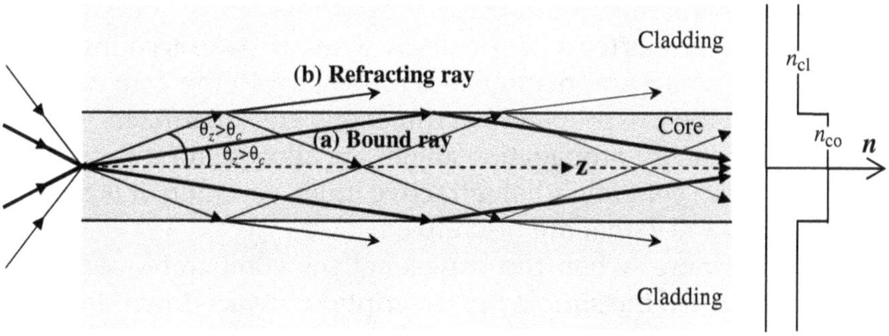

FIGURE 1.2

Ray projection on cross-section of circular step-index optical fiber for (*a*) a meridional ray path, and (*b*) a skew ray path. For step index fibers, the azimuthal angle, θ_ϕ = the angle between the projection of the path onto the fiber cross-section and the tangent to the interface (azimuthal direction), is invariant.

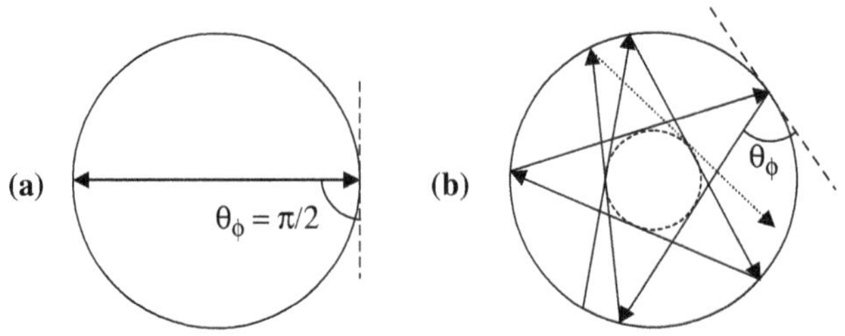

However, for rays with θ_z greater than θ_c, the rays refract with loss from the core occurring at each bounce. In the case of optical fibers, the rays shown are propagating through the fiber axis in Fig. 1.1. Optical fibers also support skew rays which follow helical paths bouncing around the fiber (rather than through the axis) as they propagate along it as shown in Fig. 1.2. For circularly symmetric, longitudinally invariant, step-index fibers, both the axial ray angle θ_z of Fig. 1.1, and the azimuthal angle, θ_ϕ of Fig. 1.2, are invariant along the fiber. When the core refractive index profile is no longer constant, but varies with radial distance r from the core center of a circularly symmetric fiber (as is the case for a graded index fiber), then the ray angles θ_z and θ_ϕ vary with position. However, longitudinal invariance leads to the axial invariant $\bar{\beta} = n_{\text{eff}} = n(r) \times \cos\theta_z(r)$, and circular symmetry (azimuthal invariance) of the fiber leads to the azimuthal invariant $\bar{l} = (r/\rho) \times n(r) \sin\theta_z(r) \cos\theta_\phi(r)$. These invariants are analogous to energy and angular momentum respectively in classical mechanics [93].

For small cores (or more precisely when we take account of the wavelength being non-negligible with respect to the core radius), we find that the ray propagation directions are "quantized"—only certain discrete ray propagation angles are allowed for a step index fiber. For more general radial refractive index variation, it is the ray invariants $\bar{\beta}$ and $\bar{l}$ that are discretized.

Furthermore, when the wavelength is comparable with the core dimensions, the simple ray description breaks down. Instead, propagation must be described in terms of electromagnetic waves which may be decomposed in terms of electromagnetic modes with each mode having a dominant local ray propagation angle (or more precisely being constructible in terms of plane waves of a range of angles with the dominant angle being that of the associated ray). For an in-depth ray optics description, we refer to Part I of Ref. 3.

1.5.2 Wave Optics Description

Field Dependence in Propagation Direction (Longitudinal Dependence)

In a medium of uniform refractive index, n, an (infinite) electromagnetic plane wave, that is monochromatic with frequency ω and propagating in the z-direction, has z-dependence $\cos[knz - \omega t] = \mathrm{Re}\{\exp[i(knz - \omega t)]\}$, which is usually simply written in complex exponential form as $\exp[i(knz - \omega t)]$, where $k = 2\pi/\lambda = \omega/c$ is the free-space wave number, and λ, the free-space wavelength, and $v_p = c/n$ its phase velocity. This corresponds to a Fourier component of a wave that is finite in extent and given by an integral over all frequencies.

If that wave now propagates in a z-directed and z-invariant waveguide such as that shown in Fig. 1.1, then its field may be decomposed in terms of modes having field components, each of which is separable as a product of transverse (x, y) and longitudinal (z) dependencies. The longitudinal dependence is now $\exp[i(kn_{\mathrm{eff}}z - \omega t)]$, and the phase velocity $v_p = c/n_{\mathrm{eff}}$, where the effective refractive index, n_{eff}, that the mode sees is between the core and cladding refractive indices, n_{co} and n_{cl} respectively, for bound or guided waves. One can think of the value of n_{eff} associated with a particular mode as being a weighted "average" refractive index that the mode "sees" whose value depends on how much of the mode is in the core, and how much is in the cladding. Furthermore, one finds that there are only certain discrete values of n_{eff} that allow the field to satisfy the appropriate boundary conditions at the core-cladding interface and to decay to zero, far from the core as is required for completely guided or bound waves.

Drum Mode Analogy with Transverse Field Dependence

For a circular fiber, the allowed bound modes have transverse electromagnetic field amplitude forms [as will be seen in the scalar mode forms $\psi(r, \phi)$ of Fig. 3.1] that are somewhat analogous to

FIGURE 1.3

Lowest-order drum modes (*l*, *m*) where *l* is the number of zero lines in the azimuthal direction and *m* is the number of nodes from the center in the radial direction. Note that rotations of (1, 1) by $\pi/2$ and (2, 1) by $\pi/4$ are also modes.

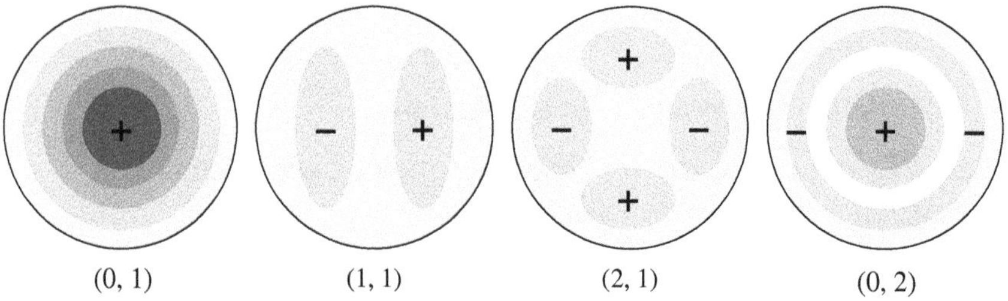

(0, 1) (1, 1) (2, 1) (0, 2)

the membrane displacement amplitude forms for the modes of a drum[1] shown schematically in Fig. 1.3, particularly in the case of strongly guided modes for which the field is essentially zero at the core-cladding interface, just as the membrane stretched over the top of a drum is constrained to have zero amplitude at the rim of the drum.

Quantum-Mechanical Analogy

As described in Fig. 1.2, and discussed in more detail in Ref. 93 (see also Refs. 141–142), a waveguide refractive index profile is analogous to an "upside-down" potential well in quantum mechanics. As we will find in Chap. 2, for a longitudinally (z) invariant, optical fiber, the transverse electromagnetic field, $\psi(\mathbf{r})$, where $\mathbf{r} = (x, y)$ or (r, ϕ), satisfies the scalar wave equation:

$$\left\{\nabla^2 + \frac{2\Delta n_{co}^2}{\lambda\!\!\!^{-2}}\left[\frac{U^2}{V^2} - f(\mathbf{r})\right]\right\}\psi(\mathbf{r}) = 0 \qquad (1.1)$$

where $\lambda\!\!\!^{-} \equiv \lambda/2\pi \equiv 1/k$ and the other notation is given in Table 2.1. In particular, the normalized modal eigenvalue (normalized transverse propagation constant) $U = \rho k(n_{co}^2 - n_{eff}^2)^{1/2}$, guidance parameter (normalized frequency) $V = \rho k(n_{co}^2 - n_{cl}^2)^{1/2}$ and normalized profile height parameter $\Delta = [1 - (n_{cl}/n_{co})^2]^{1/2}$. This may be compared with the Schrödinger equation for a particle

[1] http://en.wikipedia.org/wiki/Vibrations_of_a_circular_drum

with unit mass and energy E in a time-independent potential well $\upsilon(\mathbf{r})$:

$$\left\{\nabla^2 + \frac{2}{\hbar^2}\left[E - \upsilon(\mathbf{r})\right]\right\}\Psi(\mathbf{r}) = 0, \tag{1.2}$$

where $\hbar \equiv h/2\pi$. We see that Planck's constant h plays a role corresponding to wavelength λ; mechanical energy E is related to the square of the normalized propagation constant U:

$$E \leftrightarrow \Delta n_{co}^2 U^2/\upsilon^2 = \frac{1}{2}\left[n_{co}^2 - n_{eff}^2\right], \tag{1.3}$$

and the mechanical potential corresponds to the normalized refractive index profile

$$\upsilon(\mathbf{r}) \leftrightarrow \Delta n_{co}^2 f(\mathbf{r}) = \frac{1}{2}\left[n_{co}^2 - n^2(\mathbf{r})\right] \tag{1.4}$$

Figure 1.4 provides example of modal effective index spectra analogous to energy level spectra. Figure 1.5 provides the graphical

FIGURE 1.4

Modal spectra: (*a*) Bound modes have a discrete spectral levels (- - -) with $n_{cl} < n_{eff}$ ($\equiv \beta/k$) $< n_{co}$, and radiation modes have a continuous spectrum with $n_{eff} < n_{cl}$. Note that radiation modes may be divided into (i) propagating radiation modes with $0 < n_{eff} < n_{cl}$, and (ii) evanescent radiation modes with n_{eff} imaginary, that is $\beta^2 \equiv (kn_{eff})^2 < 0$, which thus decay as exp $(-\beta^i z)$ rather than oscillate as Re $\{\exp(-i\beta^r z)\} = \cos(\beta^r z)$ in the propagation direction. (*b*) The "upside-down" normalization giving $0 < U < V$ for bound modes and $U < V$ for radiation modes allows a comparison with quantum-mechanical energy level spectra. The value of U and number of bound modes depends on the value of V. Here we show the case of two bound modes.

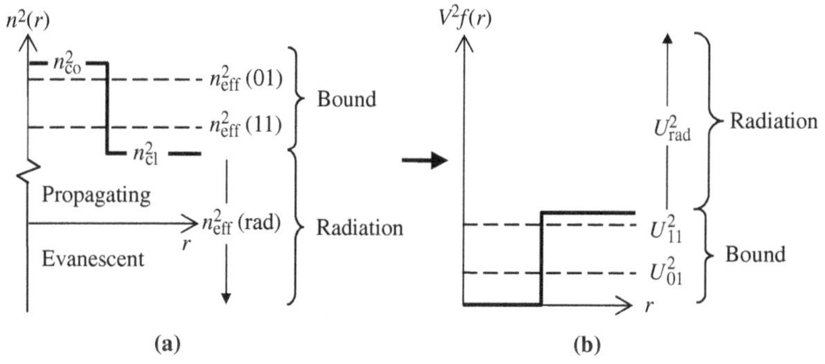

FIGURE 1.5

Relationship of normalized transverse propagation constant or modal core parameter (eigenvalue) U to the longitudinal propagation constant β.

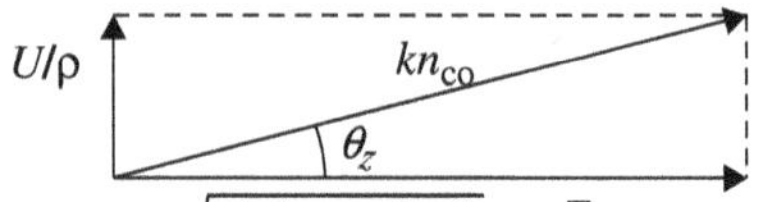

$$\beta \equiv kn_{\text{eff}} = \sqrt{k^2 n_{\text{co}}^2 - U^2/\rho^2} \rightarrow k\bar{\beta} \equiv kn_{\text{co}}\cos\theta_z$$

relationship between n_{eff} and U, and Fig. 1.6 shows the dependence of U on V for a circular step index fiber under the assumption of an infinite cladding.

Figure 1.5 shows the relationship between U and β. Note that for large V it may be shown [3, 93] that (1) the propagation constant is related to the axial invariant as $\beta \equiv kn_{\text{eff}} \rightarrow k\bar{\beta} \equiv kn_{\text{co}}\cos\theta_z$, and (2) the azimuthal mode number (analogous to the angular momentum quantum number) is related to the azimuthal invariant as $l \rightarrow \rho k\bar{l} = r\,n(r)\,\sin\theta_z(r)\,\cos\theta_\phi(r)$.

FIGURE 1.6

Normalized modal eigenvalues $U_{lm} = U_{lm}(V)$ for the first four mode forms of a circular step-index fiber within the weak guidance approximation obtained by requiring continuity of the scalar field and its first derivative at the core-cladding interface.

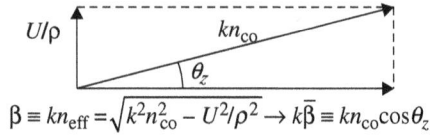

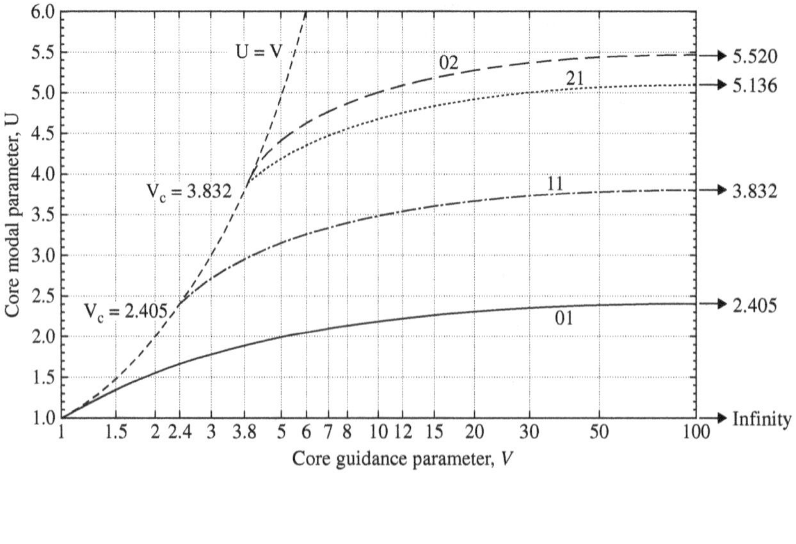

Finite-Cladding Fibers

Single-mode optical fibers require guidance parameter V less than second mode cutoff ($V_c \approx 2.4$ for step-index fiber as in Fig. 1.6), and typically have a core diameter $\phi_{co} \equiv 2\rho$ on the order of 10 μm and a cladding diameter $\phi_{cl} \equiv 2\rho_{cl}$ of 125 μm (= 1/8 mm) as shown in Fig. 1.7(a). As a first approximation, the cladding is treated as infinite and those electromagnetic modes that are not bound to the core as part of the radiation continuum as in Fig. 1.4. However, when one considers the finite nature of the cladding [97, 143, 144], strictly speaking, (1) the continuum of radiation modes associated with the cladding is discretized in terms of cladding modes as shown in Fig. 1.7(b), and (2) the fundamental (01) "core" mode can transition to being guided by the cladding with $n_{eff} < n_{cl}$ when V is

FIGURE 1.7

Finite-cladding fiber [97, 143–144] with a core, cladding and external (air or jacket) refractive indices n_{co}, n_{cl} and n_{ex} respectively: (*a*) fiber cross-section, (*b*) refractive index profile —— and modal effective indices – – –.

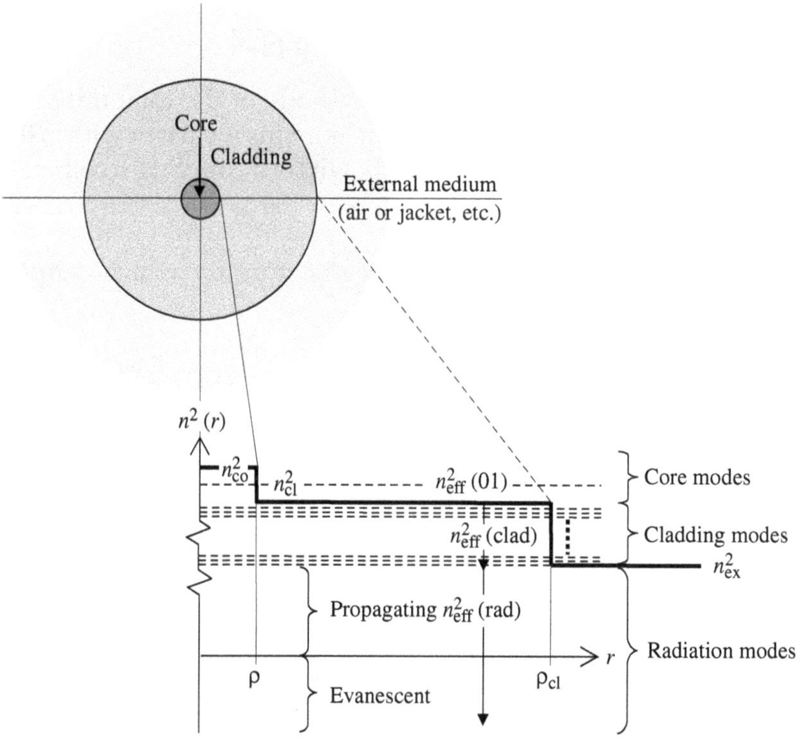

small enough, that is, when below V is what is sometimes referred to as core-mode "cutoff" [97], $V_{cc} \approx \sqrt{2/\ln(\rho_{cl}/\rho)}$.

Thus, so called "single-mode" fibers, are not really single-moded; in addition to the fundamental "core" mode, they can support several thousand cladding modes with closely spaced values of n_{eff} (which are sometimes approximated as a continuum of radiation modes if the reflection from the cladding external medium interface is not of physical importance). Optical fibers used for transmission typically have a lossy jacket that rapidly strips off the cladding modes so that only the fundamental (core) modes propagates over any significant distance. However, there are some cases when it is convenient or necessary to take account of these discrete cladding modes such as in tapered "single-mode" fibers for which coupling can occur between the fundamental "core" mode and the higher order "cladding" modes.

1.5.3 Adiabatic Transitions and Coupling

Straight Waveguide

A mode of a longitudinally invariant waveguide has transverse form which is also longitudinally invariant

$$\Psi(x, y, z) = \psi(x, y) \exp [i\beta z] \tag{1.5}$$

where β = invariant. If that mode is a mode of the total transverse structure and that transverse structure remains invariant with z, then that mode propagates independently without coupling to others.

Coupling Structures

Figure 1.8 considers three classes of structure for which coupling between waveguide modes can occur.

Tapers and Butt Joints

In the case of Fig. 1.8(a), we may consider three cases:

1. **Slowly varying (adiabatically) tapered waveguide:** The concept of a straight waveguide mode may be extended to longitudinally varying waveguides using local modes [3, 120], that is, the modes of a waveguide that is approximately straight locally. If the variation is "slow," then

$$\Psi(x, y, z) = a\hat{\psi}(x, y, n(z)) \exp\left[\int_0^z i\beta(z')dz'\right], \tag{1.6}$$

$$\beta(z) = \beta(n(z)), \text{ adiabatic approximation}$$

FIGURE 1.8

Coupling structures: (*a*) Tapered finite-cladding optical fiber: For "slow" tapering, a modes evolves adiabatically along the taper. For faster tapers, the mode couples to modes of the same symmetry. (*b*) Butt joint between waveguides of different diameters supporting modes of different sizes. Coupling between the modes of the input guide and the output guide is determined by an overlap integral, so that the total field remains constant across the joint. (*c*) Parallel waveguides: excitation of a mode of one of the waveguides corresponds to excitation of two supermodes of the two-waveguide structure. Beating of these supermodes results in coupling of light from one waveguide to the other.

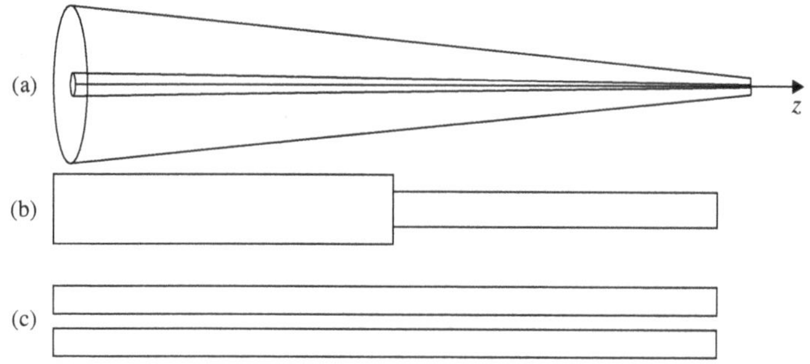

where $\hat{\psi}$ is modal field solution of the locally straight waveguide with index $n(x, y)$ at position z along the waveguide. The caret on ψ indicates normalization so that its norm integrated over the cross section gives unit power.

2. **Fast taper:** A fast taper between waveguides of two different sizes occurs when the taper length is smaller than the beat length $z_b = 2\pi/(\beta_1 - \beta_2)$ between the closest modes of the same symmetry. Such a taper may be approximated as a butt joint [as in the "sudden approximation"; Fig. 1.8(b)].
3. **Intermediate taper:** In general, a taper may be approximated as a concatenation of many butt joints, that is, a staircase with coupling between modes occurring along a significant portion of the taper. More precisely, the field is given as a summation of forward- and backward-propagating modes whose field within the scalar approximation is given by

$$\Psi(x, y, z) = \sum_j \left[b_j(z) + b_{-j}(z) \right] \hat{\psi}_j(x, y, n(z)) \tag{1.7}$$

where the modal field amplitudes along the taper $b_j(z)$ and $b_{-j}(z)$, for the forward and backward propagating modes respectively, are in general determined numerically [3, 32, 120, 145], e.g., as solutions of coupled local mode equations.

Coupled Parallel Waveguides

In the case of Fig. 1.8(c), although the total structure is longitudinally invariant, excitation of the fundamental mode of one of the guides, i.e., a substructure mode, corresponds to excitation of two normal modes of the total structure, i.e., two *supermodes*, each of which have different phase velocities and thus beat along the two-waveguide structure resulting in oscillation of light between the two waveguides. Determination of the supermodes of multiwave guide or multicore structures is the subject of Chap. 6.

1.6 OUTLINE AND MAJOR RESULTS

The material in the remainder of the book is organized as below.

- Chapter 2 develops the basic wave equations and parameters using the weak guidance formalism for longitudinally invariant optical waveguides.
- Chapters 3 to 5 discuss single-core fibers (circular, $\mathbf{C}_{nv}$ symmetric and anisotropic) and lay the foundation for analysis of multicore fibers in Chap. 6.
- Chapter 7 discusses longitudinal variations and recent developments.

In more detail, the major themes and result highlights are as follows.

- In Chap. 3, we illustrate the consequences of symmetry using a group theoretic approach in degeneracy determination and field construction of different modes of circularly symmetric fibers (including the circularly polarized (CP) representation).
    - For each scalar mode ψ_{01}, ψ_{11} (even and odd), ψ_{21} (even and odd), ψ_{02}, etc., there are two linearly polarized pseudo vector modes: x and y polarized LP_{01}, LP_{11} (even and odd), LP_{21} (even and odd), LP_{02}, etc.
    - Symmetry shows that when coupling of field components is considered, the correspondence with the true vector modes is $LP_{01} \rightarrow HE_{11}$, $LP_{11} \rightarrow TE_{01}$, TM_{01}, HE_{21}, $LP_{21} \rightarrow EH_{11}$, HE_{31}, etc., where HE and EH modes are hybrid modes with nonzero longitudinal field components,

TE (transverse electric) modes have zero longitudinal electric field components and TM (transverse magnetic) modes.
- Note that for a uniform core, when the dielectric cladding is replaced by a metal cladding (i.e., for circularly symmetric metal waveguides filled with a homogeneous dielectrics) all modes are TE or TM, and we have the replacements $HE_{11} \rightarrow TE_{11}$, $HE_{21} \rightarrow TE_{21}$, $EH_{11} \rightarrow TM_{11}$ [74, 146–147], etc.

- In Chap. 4, we consider the degeneracy splitting and modal field transformation resulting when the *azimuthal* circular symmetry is lowered to n-fold rotation-reflection symmetry $\mathbf{C}_{nv}$ as illustrated by distortion of the cross-section of a fiber from circular to elliptical, triangular, square, pentagonal and hexagonal (n = 2, 3, 4, 5 and 6 respectively).
  - The two polarizations of the fundamental mode are split for the case $n = 2$ (e.g., elliptical perturbations), but remain degenerate for 3, 4, 5 and 6.
  - The two polarizations of the HE_{21} mode (and higher order modes) are also split for the case $n = 2$, but remain degenerate for 3, 4, 5 and 6. Furthermore all modes of the second mode set (TE_{01}, TM_{01} and HE_{21}) transition to LP modes as an elliptical perturbation is strengthened.
  - The two polarizations of the HE_{31} mode are split for $n = 3$ (triangular) perturbations which from a symmetry standpoint has the same consequences as $n = 6$ (hexagonal) perturbations.
- In Chap. 5, we consider linearly, radially (or azimuthally) and circularly birefringent (gyrotropic) fibers.
  - We develop the wave equations for these anisotropies.
  - We provide qualitative level splitting and mode form transition results depending on competing strengths of the refractive index profile height and each of the anisotropies.
- Chapter 6 is devoted to the construction of modes of multicore fibers or multifiber coupler arrays with discrete rotation symmetry $\mathbf{C}_n$ or rotation-reflection symmetry $\mathbf{C}_{nv}$.
  - Using general basis functions applicable to arrays of few-mode cores, we first obtain approximations to the scalar and vector supermode field forms as simple combinations of the isolated core fields together with propagation constant degeneracies.

  - We consider symmetry consequences regarding the field form dependence on the relative magnitudes of (1) the core-separation related intercore coupling and (2) the index-profile-height-related polarization coupling.
  - Using these results as a basis, we then discuss how symmetry can simplify numerical field evaluation and give quantitative determination of propagation constant splittings.
  - Finally, we consider example applications of the supermode forms and propagation constant splittings to determine the power transfer and polarization rotation characteristics of multicore fibers or couplers.
- Chapter 7 provides a summary of the results and discusses extensions of the concept of modes for longitudinally invariant structures to modes for structures with longitudinal variations, such as periodic structures, and Kerr-type nonlinear waveguides where intensity-dependence induces longitudinal variation in the presence of a propagating wave. Sections 7.4 through 7.8 consider recent development, particularly in waveguide modeling software, photonic crystals and quasi crystals, photonic crystal fibers, and fiber Bragg gratings (FBGs).
- The reader is referred to the Appendix and to Refs. 6 and 9 for group theoretical definitions.

Throughout the text, details that are interesting and useful for reference but nonessential for the main development are indicated before and after by the section sign (§). We suggest that these be passed over on a first reading.

CHAPTER 2

Electromagnetic Theory for Anisotropic Media and Weak Guidance for Longitudinally Invariant Fibers

In this chapter first we define the electromagnetic media to be considered; then we develop the basic wave equations for electrically anisotropic and isotropic media, considering step by step the simplifications resulting from special forms of the dielectric tensor. Finally we review the aspects applicable to this chapter of the weak-guidance formalism of Snyder and coworkers [3, 48] as applied to *longitudinally invariant* isotropic fibers, leaving further discussion of weak guidance for anisotropic fibers to Chap. 5. The waveguide and coordinate notation to be used is summarized in Table 2.1.

2.1 ELECTRICALLY ANISOTROPIC (AND ISOTROPIC) MEDIA

We consider source-free media of *uniform magnetic permeability* $\mu = \mu_0$, magnetic flux $\mathbf{B} = \mu_0\mathbf{H}$, and *electrical anisotropy* with electric polarization manifest via the electric permittivity $\mathbb{\varepsilon} = \varepsilon_0 \mathbb{n}^2$, giving the displacement electric field $\mathbf{D} \equiv \varepsilon_0 \mathbb{n}^2\, \mathbf{E}$.

TABLE 2.1

Notation

(*a*) Coordinate Notation	
$\mathbf{r} = \mathbf{r}_t = (x, y) = x\hat{\mathbf{x}} + y\hat{\mathbf{y}}$ and $(r, \phi) = r\hat{\mathbf{r}} + \phi\hat{\boldsymbol{\phi}}$; $R = r/\rho$	Transverse cartesian and polar coordinates
$\mathbf{r} = (r_+, r_-) = r_+\hat{\mathbf{R}} + r_-\hat{\mathbf{L}},\ r_\pm = \frac{1}{\sqrt{2}}\{x + iy\} = \frac{r}{\sqrt{2}}e^{\pm i\phi}$	Circularly polarized (complex helical) coordinates
$\hat{\mathbf{R}} \equiv \{\hat{\mathbf{x}} - i\hat{\mathbf{y}}\}/\sqrt{2};\ \hat{\mathbf{L}} \equiv \{\hat{\mathbf{x}} + i\hat{\mathbf{y}}\}/\sqrt{2}$	Circularly polarized unit vectors
$\nabla \equiv \nabla_t + \hat{\mathbf{z}}\frac{\partial}{\partial z};\ \nabla_t = \hat{\mathbf{x}}\frac{\partial}{\partial x} + \hat{\mathbf{y}}\frac{\partial}{\partial y} = \hat{\mathbf{r}}\frac{\partial}{\partial r} + \frac{\hat{\boldsymbol{\phi}}}{r}\frac{\partial}{\partial \phi}$	Three- and two-dimensional (transverse) gradient operator
$\nabla^2 \equiv \nabla_t^2 + \frac{\partial^2}{\partial z^2};\nabla_t^2 = \frac{\partial^2}{\partial x^2} + \frac{\partial^2}{\partial y^2} = \frac{\partial^2}{\partial r^2} + \frac{1}{r}\frac{\partial}{\partial r} + \frac{1}{r^2}\frac{\partial^2}{\partial \phi^2}$	Three- and two-dimensional (transverse) scalar Laplacian
(*b*) Fiber Parameters	
n_{co} = maximum core index	$n^2(\mathbf{r}) = n_{co}^2\,\{1 - 2\Delta\, f(\mathbf{r})\}$ = refractive index squared
n_{cl} = cladding index	
ρ = fiber radius or scaling distance	$\Delta = \frac{1}{2}\left(1 - n_{co}^2/n_{cl}^2\right)$ = profile height parameter
λ = free-space wavelength	$k = 2\pi/\lambda = \omega/c$ = free-space wavenumber
ω = angular frequency	c = speed of light in vacuum
$V = k\rho\left(n_{co}^2 - n_{cl}^2\right)^{1/2} = k\rho n_{co}\sqrt{2\Delta}$ = guidance parameter = $\omega\rho n_{co}\sqrt{2\Delta / c}$ = normalized frequency	
(*c*) Modal Field Notation and Parameters	
$\Psi(\mathbf{r}, z, t) = \psi(\mathbf{r})e^{i(\beta z - \omega t)}$	Scalar-mode field dependence
$\mathbf{E}(\mathbf{r}, z, t) = \{\mathbf{E}_t + \hat{\mathbf{z}}E_z\}e^{-i\omega t} = \{\mathbf{e}_t(\mathbf{r}) + e_z(\mathbf{r})\hat{\mathbf{z}}\}\ e^{i(\beta z - \omega t)}$	Vector-mode field dependence
$\mathbf{e}_t = e_x\hat{\mathbf{x}} + e_y\hat{\mathbf{y}} = e_+\mathbf{R} + e_-\hat{\boldsymbol{L}}$ $e_\pm = \frac{1}{\sqrt{2}}\{e_x \pm ie_y\}$ and similarly for $\mathbf{E}_t$	Transverse field decomposition
$\beta = kn_{eff}$ = propagation constant	$n_{eff} = \beta/k$ = effective mode index
$U = k\rho\left(n_{co}^2 - n_{eff}^2\right)^{1/2}; W = k\rho\left(n_{eff}^2 - n_{cl}^2\right)^{1/2}$ = normalized core (U) and cladding (W) transverse propagation constants	

TABLE 2.1

Notation (*Continued*)

(***d***) **Mathematical Symbols**
$\forall$ = "for all" $\in$ = "is a member of the set"
Matrices are indicated by a bold open-space character, such as $\mathbb{M}$
$\oplus$ indicates direct sum; $\otimes$ indicates direct product—definition for matrices in, e.g., Ref. 75 (pp. 164 and 206)—see also Appendix for application to matrix representations
$\supset$ indicates subgroup; $\mathbf{G} \supset \mathbf{G_s}$ (or $\mathbf{G_s} \subset \mathbf{G}$) is reduction of group **G** with respect to subgroup $\mathbf{G_s}$
$\rightarrow$ indicates representation branching rule. See Sec. A.3.2, e.g., $\mathbf{M(G)} \rightarrow \mathbf{N(G_s)} \oplus \cdots$: branching of representations **M** of group **G** to irreducible representations (irreps) **N** etc. of subgroup $\mathbf{G_s}$
For other group theoretic notation, see Appendix.

In general, the refractive index squared $\mathbb{n}^2$ is a nine-component tensor which may be represented in either dyadic form (e.g., as in Ref. 74) or, as here, matrix form. (We denote matrices by bold open-space characters.)

While our focus in this chapter is almost exclusively on *electrical anisotropy*, it is worth noting in the context of both symmetry and recipes for future exotic waveguide design that *electrical anisotropy* is in fact quite a special case of more general *bianisotropy* [76] for which the electric and magnetic fields **E** and **H** are coupled via the constitutive relations $\mathbf{D} = \mathbb{\varepsilon}\mathbf{E} + \mathbb{\xi}\mathbf{H}$ and $\mathbf{B} = \mathbb{\zeta}\mathbf{E} + \mathbb{\mu}\mathbf{H}$ whose symmetry properties where originally examined by Tellegan [77]. Of particular waveguide interest is the case of chiral media [78] (embracing optical activity [76 (p. 79)] and circular dichroism [79]) for which in the bianisotropic case $\mathbb{\xi} = -\mathbb{\zeta} \equiv i\mathbb{\xi}_c$. The most studied special case of nonzero coupling terms $\mathbb{\xi}$ and $\mathbb{\zeta}$ occurs when the four tensors reduce to scalars, that is, $\mathbf{D} = \varepsilon\mathbf{E} + \xi\mathbf{H}$ and $\mathbf{B} = \zeta\mathbf{E} + \mu\mathbf{H}$ to give *bi-isotropy*, and, in particular, the case of isotropic chirality [80] ξ_c given by $\xi = -\zeta \equiv i\xi_c$, for example, for chirowaveguides [81]. Herein (Sec. 5.3) we include optical activity [82 (Chap. 6), 83, 84] in our treatment of circular birefringence via an effective (direction-dependent) dielectric tensor.

Starting from the case of general *electrical anisotropy*, in this chapter we concentrate on the following cases of simplification for the dielectric tensor:

1. Anisotropic media with a z**-aligned principal axis of refraction** such that

$$\mathbf{D}/\varepsilon_0 \equiv \mathbb{n}^2\mathbf{E} = \mathbb{n}_{\mathbb{z}}^2\mathbf{E} \equiv \mathbb{n}_{\mathbb{t}}^2\mathbf{E}_t + n_z^2 E_z \hat{z} \tag{2.1a}$$

These have the following as a particular case.

2. **"Diagonal" anisotropic media** with all three principal axes aligned with an appropriate coordinate system $(\mathbf{r}, z) = (r_1, r_2, z)$, for example, $\mathbf{r} = (r_1, r_2) = (x, y)$, (r_+, r_-) or (r, ϕ) corresponding to linear, circular, and radial birefringence discussed in Chap. 5. These media have

$$\mathbf{D}/\varepsilon_0 \equiv \mathbb{n}^2\mathbf{E} = \mathbb{n}_{\mathbb{d}}^2\mathbf{E} \equiv \mathbb{n}_{\mathbb{td}}^2\mathbf{E}_t + n_z^2 E_z \hat{z}$$
$$\equiv n_1^2\mathbf{E}_1 + n_2^2\mathbf{E}_2 + n_z^2 E_z \hat{z} \tag{2.1b}$$

with (1) $n_1 \neq n_2 \neq n_z$ corresponding to *biaxial* media, (2) $n_1 = n_2 \equiv n_t \neq n_z$ corresponding to *uniaxial* media, and as a further special case (3) $n_1 = n_2 = n_z \equiv n$,

3. **Isotropic media** for which $\mathbb{n}^2$ is replaced by the scalar n^2, that is,

$$\mathbf{D}/\varepsilon_0 \equiv \mathbb{n}^2\mathbf{E} = \mathbb{n}_{\mathbb{i}}^2\mathbf{E} \equiv n^2 E \tag{2.1c}$$

2.2 GENERAL WAVE EQUATIONS FOR ELECTRICALLY ANISOTROPIC (AND ISOTROPIC) MEDIA

For the general media above, assuming ***time harmonic fields*** $\mathbf{e}^{-i\omega t}$, **Maxwell's equations** for the spatial dependence of the electric and (displacement) magnetic fields $\mathbf{E}(\mathbf{r}, z)$ and $\mathbf{H}(\mathbf{r}, z) = \mu_0^{-1}\mathbf{B}$ in rationalized mks units take the form

$$\boxed{\begin{array}{llll} \nabla \times \mathbf{E} = ik(\varepsilon_0/\mu_0)^{1/2}\,\mathbf{H} & (a) & \nabla \bullet (\mathbb{n}^2\mathbf{E}) = 0 & (b) \\ \nabla \times \mathbf{H} = -ik(\varepsilon_0\mu_0)^{1/2}\,\mathbb{n}^2\mathbf{E} & (c) & \nabla \bullet \mathbf{H} = 0 & (d) \end{array}} \tag{2.2}$$

The wave equation for a monochromatic electric field of angular frequency $\omega = k(\varepsilon_0\mu_0)^{1/2}$ (or the Fourier component of a more general field) is then given by taking the curl of Eq. (2.2*a*), substituting from Eq. (2.2*c*), and using the identity $\nabla \times (\nabla \times \mathbf{E}) = \nabla(\nabla \bullet \mathbf{E}) - \mathbb{V}^2\,\mathbf{E}$ as

$$\boxed{\mathbb{V}^2\mathbf{E} + k^2\mathbb{n}^2\mathbf{E} = \nabla(\nabla \bullet \mathbf{E})} \quad \text{where} \quad \mathbb{V}^2\mathbf{E} = \mathbb{V}_{\mathbb{t}}^2\mathbf{E}_t + \hat{\mathbf{z}}\nabla_t^2 E_z + \frac{\partial^2\mathbf{E}}{\partial z^2} \tag{2.3a}$$

$$\text{with}\quad \nabla_t^2\mathbf{E}_t = \begin{bmatrix} \nabla_t^2 & 0 \\ 0 & \nabla_t^2 \end{bmatrix}\begin{bmatrix} E_x \\ E_y \end{bmatrix} \equiv \nabla_t^2\begin{bmatrix} E_x \\ E_y \end{bmatrix} = \nabla_t^2\begin{bmatrix} E_+ \\ E_- \end{bmatrix} = \begin{bmatrix} \nabla_t^2 - \dfrac{1}{r^2} & \dfrac{-2}{r^2}\dfrac{\partial}{\partial\phi} \\ \dfrac{2}{r^2}\dfrac{\partial}{\partial\phi} & \nabla_t^2 - \dfrac{1}{r^2} \end{bmatrix}\begin{bmatrix} E_r \\ E_\phi \end{bmatrix} \tag{2.3b}$$

Thus, in general, the electromagnetic field may be specified in terms of (1) three coupled scalar partial differential equations for the spatial components of the electric field **E** with appropriate boundary conditions and (2) the *j*th spatial component of the magnetic field **H** being given explicitly in terms of **E** via the first Maxwell equation, Eq. (2.2*a*), as

$$H_j = (i/k)(\mu_0/\varepsilon_0)^{1/2}\hat{\mathbf{j}} \cdot \nabla \times \mathbf{E} \tag{2.3c}$$

Polarization Effects The general wave equation, Eq. (2.3), reveals three possible sources.

1. In cylindrical polar coordinates, Eq. (2.3*b*) shows that ∇_t^2 couples the radial and azimuthal spatial components of the field. For cartesian and circularly polarized components, this polarization coupling effect is zero.
2. For anisotropic media, polarization dependence may be dominated via the permittivity tensor $\varepsilon_0\mathbb{n}^2$. For diagonal anisotropies, the *waveguide term* $k^2\mathbb{n}^2(\mathbf{r}, z)\mathbf{E}$ simply results in each polarization component E_i seeing a different guide $n_i(\mathbf{r}, z)$; for nondiagonal anisotropies it also couples the polarization components.
3. For isotropic media, Eq. (2.3*a*) gives the polarization dependence of wave propagation via the right-hand-side (RHS) *polarization term* $\nabla(\nabla \cdot \mathbf{E})$ which couples the field polarization components via the gradient of the refractive index as can be seen by using the identity for the Maxwell equation, Eq. (2.2*b*), $\nabla \cdot (n^2\mathbf{E}) = n^2\nabla \cdot \mathbf{E} + \mathbf{E} \cdot \nabla n^2 = 0$, to obtain

$$\nabla(\nabla \cdot \mathbf{E}) = -\nabla(n^{-2}\mathbf{E} \cdot \nabla n^2) = -\nabla(\mathbf{E} \cdot \nabla \ln n^2) \tag{2.4}$$

For the special case of diagonally anisotropic media for which $\{\mathbb{n}_d^2\}^{-1}$ simply consists of the inverse components n_i^{-2}, this convenient reexpression generalizes to

$$\nabla(\nabla \bullet \mathbf{E}) = -\nabla\left(\{\mathbb{n}_d^2\}^{-1}\mathbf{E} \bullet \nabla\{\mathbb{n}_d^2\}\right)$$

$$= -\nabla\left(\mathbf{E} \bullet \nabla \ln\{\mathbb{n}_d^2\}\right) \quad \text{where} \quad \nabla \ln\{\mathbb{n}_d^2\} = \begin{bmatrix} \nabla \ln n_1^2 \\ \nabla \ln n_2^2 \\ \nabla \ln n_z^2 \end{bmatrix} \tag{2.5}$$

We note that for media with *magnetic anisotropy* $\mu(\mathrm{r})$ but electrical isotropy and homogeneity $\varepsilon = \varepsilon_0$, the entire formalism here for **E** in the presence of electrical anisotropy may still be used for **H** by noting that one obtains an identical wave equation, Eq. (2.3*a*), upon replacement of **E** by **H** and $\mathbb{n}^2$ by μ/ε_0 with the sign difference between the Maxwell equations (2.2*a*) and (2.2*c*) only appearing in the analog of Eq. (2.3*c*) for explicit evaluation of **E** in terms of **H**.

For media that are both electrically and magnetically nonhomogeneous, reexpression analogous to Eq. (2.5) leads to an extra term $-(\nabla \ln\{\mu_d\}) \times (\nabla \times \mathbf{E})$ on the RHS of the wave equation for **E**; cf. Eqs. (1.4–7) and (1.4–8) of Ref. 85 for the isotropic case.

For chiral media a term proportional to $\nabla \times \mathbf{E}$ cannot be eliminated from the wave equation for **E** (e.g., see Ref. 80 for the homogeneous isotropic case). This reveals the asymmetry associated with, e.g., optical activity in that curl alone (i.e., a single operation involving $\times$ is not a vector under reflection of the coordinate system [86]).

2.3 TRANSLATIONAL INVARIANCE AND MODES

The first major waveguide symmetry to be considered here is that of translational invariance. For a longitudinally invariant *z*-aligned guide of refractive index $n(\mathbf{r})$, or more generally $\mathbb{n}^2(\mathbf{r})$, one may separate the field in terms of independently propagating modes with longitudinal dependence $e^{i\beta z}$ such that each has an electric field of the form

$$\boxed{\mathbf{E}(\mathbf{r}, z) = \mathbf{e}(\mathbf{r})e^{i\beta z} = \{\mathbf{e}_t(\mathbf{r}) + e_z(\mathbf{r})\hat{\mathbf{z}}\}e^{i\beta z}} \qquad \mathbf{r} = (x, y), (r_+, r_-), \text{ or } (r, \phi) \tag{2.6}$$

as in Table 2.1 where fiber parameters are defined.

We remark that there are several special lower z-dependent "symmetries" that allow a separation of the longitudinal dependence so that the field may be expressed in terms of modes: (1) periodic guides discussed in Sec. 7.2, (2) a homogeneous dielectric wedge that is metal-clad [87 (p. 366)] (or has large guidance parameter) so that the field can be taken as zero at the guide boundary and thus remaining invariant under a scaling, (3) some special graded-index parabolic tapers [88], and (4) slowly varying or "adiabatic" tapers for which translational invariance remains an approximate local symmetry and the field is (approximately) separable in terms of local modes which, to an excellent approximation, propagate independently. For nonadiabatic tapers, one needs to take account of coupling between the local modes [31, 32, 120].

2.4 WAVE EQUATIONS FOR LONGITUDINALLY INVARIANT MEDIA

2.4.1 General Anisotropic Media

Substitution of the longitudinal dependence in the basic wave equation, Eq. (2.3), then leads to the wave equation for the transverse (**r**) *dependence* of the three-component field **e**(**r**)

$$\boxed{\left\{\nabla_{t}^{2}+k^{2}\mathbb{n}^{2}(\mathbf{r})-\beta^{2}\right\}\mathbf{e}(\mathbf{r})=(\nabla_{t}+i\beta\hat{\mathbf{z}})(\nabla_{t}\cdot\mathbf{e}_{t})+i\beta(\nabla_{t}+i\beta\hat{\mathbf{z}})e_{z}} \tag{2.7}$$

However, while z invariance allows elimination of the z dependence and results in an eigenvalue problem for propagation constant β, for general anisotropies $\mathbb{n}^2$, the wave equation still couples the transverse *component* $\mathbf{e}_t$ with the longitudinal component e_z and cannot be reduced from the equivalent three coupled scalar differential equations.

For example, coupling of $\mathbf{e}_t$ with e_z by the dielectric tensor is the case for twisted fibers as discussed in Ref. 89.

2.4.2 Anisotropic Media with z-Aligned Principal Axis

For anisotropic media with a z-aligned principal axis of refraction as in Eq. (2.1*a*), *longitudinal invariance* together with the Maxwell equation $\nabla \cdot (\mathbb{n}^2\mathbf{E}) = 0$ gives the longitudinal field component explicitly in terms of the transverse components as

$$e_z(\mathbf{r})=\frac{i}{\beta n_z^2}\nabla_t\cdot(\mathbb{n}_t^2\mathbf{e}_t) \tag{2.8}$$

Thus e_z may be eliminated from the transverse component of the wave equation which reduces to

$$\boxed{\left\{\nabla_t^2 + k^2 \mathbb{n}_t^2(\mathbf{r}) - \beta^2\right\}\mathbf{e}_t(\mathbf{r}) = \nabla_t \left\{\nabla_t \bullet \mathbf{e}_t - n_z^{-2}\nabla_t \bullet (\mathbb{n}_t^2 \mathbf{e}_t)\right\}} \tag{2.9}$$

That is, the field may be determined via solution of only two coupled scalar differential equations in the transverse field components, rather than three for more general anisotropies.

2.4.3 "Diagonal" Anisotropies

Furthermore, for diagonal $\mathbb{n}^2 = \mathbb{n}_d^2$ and thus diagonal $\mathbb{n}_t^2 = \mathbb{n}_{td}^2$ in the appropriate coordinate system [as in Eq. (2.1*b*)], Eqs. (2.8) and (2.9) for the longitudinal and transverse field electric field components may be reexpressed using the diagonal matrix identities

$$\nabla_t \bullet \left(\mathbb{n}_{td}^2 \mathbf{e}_t\right) = \left\{\mathbb{n}_{td}^2 \nabla_t\right\} \bullet \mathbf{e}_t + \mathbf{e}_t \bullet \nabla_t \left\{\mathbb{n}_{td}^2\right\} \quad \text{and} \quad \left\{\mathbb{n}_{td}^2\right\}^{-1} \nabla_t \left\{\mathbb{n}_{td}^2\right\} = \nabla_t \ln \left\{\mathbb{n}_{td}^2\right\} \tag{2.10}$$

where $\nabla_t \ln \{\mathbb{n}_{td}^2\}$ is the two-dimensional analog of the term in Eq. (2.5).

In particular, this gives the wave equations of Chap. 5 where we will consider the "diagonal" anisotropies corresponding to linear, radial, and circular birefringence. These have the general form

$$\boxed{\left\{\nabla_t^2 + k^2 \mathbb{n}_{td}^2(\mathbf{r}) - \beta^2\right\}\mathbf{e}_t(\mathbf{r}) = \nabla_t \left\{(2\mathbb{\delta}_{zt} \nabla_t) \bullet \mathbf{e}_t - [(\mathbb{1} - 2\mathbb{\delta}_{zt})\mathbf{e}_t] \bullet \nabla_t \ln\left\{\mathbb{n}_{td}^2\right\}\right\}} \tag{2.11a}$$

where

$$\mathbb{n}_{td}^2 = \begin{bmatrix} n_1^2 & 0 \\ 0 & n_2^2 \end{bmatrix} \quad \text{and} \quad 2\mathbb{\delta}_{zt} = \mathbb{1} - \mathbb{n}_{td}^2 / n_z^2 = 2\begin{bmatrix} \delta_{z_1} & 0 \\ 0 & \delta_{z_2} \end{bmatrix}$$

$$\text{with } 2\delta_{zi} = 1 - n_i^2 / n_z^2 \tag{2.11b}$$

For **uniaxial** media $\mathbb{n}_{td}^2$ and $\mathbb{\delta}_{zt}$ in Ref. 2.11 are simply replaced by the scalars n_t^2 and δ_{zt}.

2.5 TRANSVERSE FIELD VECTOR WAVE EQUATION FOR ISOTROPIC MEDIA

For longitudinally invariant isotropic media, Eq. (2.8) for the longitudinal field component reduces to

$$e_z(\mathbf{r}) = \frac{i}{\beta n^2}\nabla_t \bullet (n^2 \mathbf{e}_t) = \frac{i}{\beta}\{\nabla_t \bullet e_t + e_t \bullet \nabla_t \ln n^2\} \tag{2.12}$$

From Eq. (2.11*a*) in the limit $\delta_{zt} = 0$, or simply by noting that the polarization component coupling term in the basic wave equation, Eq. (2.3), reduces to $\nabla(\nabla \cdot \mathbf{E}) = -\nabla\{\mathbf{E}_t \cdot \nabla \ln n^2(\mathbf{r})\}$, we obtain the vector wave equation (VWE) for the transverse field $\mathbf{e}_t$ in a cartesian or circularly polarized basis as [3 (Eq. 30-18)]

$$\begin{aligned}&\left\{\nabla_t^2 + k^2 n^2(\mathbf{r}) - \beta^2\right\}\mathbf{e}_t(\mathbf{r}) = -\nabla_t\{\mathbf{e}_t \bullet \nabla_t \ln\, n^2(\mathbf{r})\}:\\ &\mathbf{r} = (x, y) \text{ or } (r_+, r_-)\text{: VWE}\end{aligned} \tag{2.13}$$

where parameters are defined in Table 2.1 and for cylindrical polar coordinates $\mathbf{r} = (r, \phi)$, ∇_t^2 is replaced by $\mathbb{V}_t^2$ of Eq. (2.3*b*).

Isotropic media polarization effects are now manifest via the RHS term

$$\mathbb{H}_{\text{pol}}\, \mathbf{e}_t \equiv -\nabla_t\{\mathbf{e}_t \bullet \nabla_t \ln\, n^2(\mathbf{r})\} = -\nabla_t\left\{e_t \bullet \frac{\nabla_t n^2}{n^2}\right\} \tag{2.14}$$

which couples the transverse field polarization components.

2.6 SCALAR WAVE EQUATION

If we neglect the term $\mathbb{H}_{\text{pol}}\mathbf{e}_t$ and then decompose $\mathbf{e}_t$ into two linearly (or circularly) polarized components e_x and e_y (or e_+ and e_-), these are uncoupled and for *isotropic* media each component satisfies the *same* scalar wave equation (SWE)

$$\begin{aligned}&\left\{\nabla_t^2 + k^2 n^2 - \tilde{\beta}^2\right\}\psi(\mathbf{r}) = 0 \qquad \text{i.e.,} \qquad \mathcal{H}_s \psi = \tilde{\beta}^2 \psi\\ &\text{with} \qquad \mathcal{H}_s = \nabla_t^2 + k^2 n^2(\mathbf{r}) \qquad \text{SWE}\end{aligned} \tag{2.15}$$

Thus neglect of $\mathbb{H}_{\text{pol}}\mathbf{e}_t$ leads to polarization-independent propagation, and we can choose linearly polarized [3 (Secs. 13-5 and 13-7)]

(LP) modes, or circularly polarized (CP) modes (as well as arbitrary linear combinations thereof), as solutions of the corresponding vector wave equation. However, consideration of $\mathbb{H}_{\mathbb{pol}}\mathbf{e}_t$ leads to a specific mixing of polarization components as discussed in the next subsection.

We remark that for media that are "diagonally" anisotropic in a linearly polarized (cartesian) or circularly polarized basis (i.e., linear or circular birefringence), similar neglect of the RHS of the corresponding VWE, Eq. (2.11), leads to a SWE for each component e_i with n^2 being replaced by n_i^2. However, in contrast to the isotropic case, for $n_1^2 \neq n_2^2$, each component e_i satisfies a *different* SWE with different solutions $\tilde{\beta}_1^2$ and $\tilde{\beta}_2^2$. Thus although the VWE without the RHS will lead to LP (or CP) modes for linear (or circular) birefringence, their nondegeneracy means that we cannot take linear combinations of them. This will be further discussed in a symmetry context in Chap. 5. For the rest of this chapter we concentrate on developments for isotropic guides.

2.7 WEAK-GUIDANCE EXPANSION FOR ISOTROPIC MEDIA

When the profile variations are small, incorporation of polarization effects is best seen in using a standard perturbation theory construction of the field [3, 48, 69]. In particular, the weak-guidance approach considers a perturbation expansion in terms of the profile height parameter Δ of Table 2.1, giving

$$\nabla_t \ln n^2(\mathbf{r}) = -\sum_{n=1}^{\infty} \Delta^n (2/n)^n \ \nabla_t (f^n) \approx -2\Delta \nabla_t f \qquad \text{for } \Delta << 1 \quad (2.16a)$$

with

$$\mathbf{e}_t = \tilde{\mathbf{e}}_t + \sum_{n=1}^{\infty} \Delta^n \mathbf{e}_t^{(n)} \approx \tilde{\mathbf{e}}_t + \Delta \mathbf{e}_t^{(1)} \quad (2.16b)$$

$$U = \tilde{U} + \sum_{n=1}^{\infty} \Delta^n U^{(n)} \approx \tilde{U} + \Delta U^{(1)} \quad (2.16c)$$

where $\tilde{\mathbf{e}}_t$ and $\tilde{U} = \rho(k^2 n_{co}^2 - \tilde{\beta}^2)^{1/2}$ are, respectively, eigenvectors and eigenvalues of the VWE to zeroth order in Δ, denoted VWE$^{(0)}$. This is the *polarization–independent* VWE without $\mathbb{H}_{pol}\mathbf{e}_t$, which, in terms of normalized parameters of Table 2.1, is

$$\left\{\rho^2\nabla_t^2+\tilde{U}^2-V^2f(\mathbf{r})\right\}\tilde{\mathbf{e}}_t(\mathbf{r})=0 \quad \text{or} \quad \mathbb{H}_0\,\tilde{\mathbf{e}}_t=\tilde{U}^2\mathbf{e}_t \quad \mathrm{VWE}^{(0)} \tag{2.17a}$$

where

$$\mathbb{H}_0\equiv\mathcal{H}_0\mathbb{1}\equiv\begin{bmatrix}\mathcal{H}_0 & 0\\ 0 & \mathcal{H}_0\end{bmatrix} \quad \text{and} \quad \mathcal{H}_0=-\rho^2\nabla_t^2+V^2f(\mathbf{r})\;\; \tilde{\mathbf{e}}_t=\begin{bmatrix}\tilde{e}_x\\ \tilde{e}_y\end{bmatrix} \text{ or } \begin{bmatrix}\tilde{e}_+\\ \tilde{e}_-\end{bmatrix} \tag{2.17b}$$

with $\mathbb{1}$ = unit matrix and $\mathcal{H}_0$ corresponding to the normalized SWE

$$\left\{\rho^2\nabla_t^2+\tilde{U}^2-V^2\,f(\mathbf{r})\right\}\,\psi(\mathbf{r})=0 \quad \text{i.e.,} \quad \mathcal{H}_0\,\psi=\tilde{U}^2\,\psi \quad \mathrm{SWE_n} \tag{2.18}$$

That $\mathrm{VWE}^{(0)}$ is diagonal in both cartesian and circularly polarized coordinates means that LP modes (and CP modes) are solutions; that it is a multiple of $\mathbb{1}$ (not the case for linearly or circularly polarized media considered in Chap. 5) results in arbitrary linear combinations thereof also being solutions.

However, *polarization coupling* is seen if we consider the **VWE to first order:**

$$\{\mathbb{H}_0+\Delta\mathbb{H}_{p1}\}\mathbf{e}_t=U^2\mathbf{e}_t+O(\Delta^2) \quad \text{where} \quad \mathbb{H}_{p1}\mathbf{e}_t=2\rho^2\nabla_t\{\mathbf{e}_t\boldsymbol{\cdot}\nabla_t f\} \tag{2.19}$$

which results in $\mathrm{VWE}^{(1)}$

$$\mathbb{H}_0\;\mathbf{e}_t^{(1)}=\mathbb{H}_{p1}\;\tilde{\mathbf{e}}_t-2\;\tilde{U}U^{(1)}\;\tilde{\mathbf{e}}_t \qquad \mathbf{VWE^{(1)}} \tag{2.20}$$

We note that $\mathbb{H}_{p1}$ couples the polarization components of the *zeroth-order* field $\tilde{\mathbf{e}}_t$. The result is that even in the zeroth-order weak-guidance limit, a true mode should in general consist of a mixing of both polarized components; i.e., if $\tilde{\mathbf{e}}_t$ is the zeroth-order term in the expansion (2.16*b*) of the exact field as well as a solution of $\mathrm{VWE}^{(0)}$, then only specific combinations of polarization components are possible. The standard degenerate perturbation theory recipe [90 (Sec. 31), 91 (Sec. 10-3)] for finding such combinations involves diagonalization of $\mathbb{H}_{p1}$. However, as we will show in the following sections, the appropriate combination can often be determined simply through knowledge of the symmetry of $\mathbb{H}_{p1}$ rather than its details.

2.8 POLARIZATION-DEPENDENT MODE SPLITTING AND FIELD CORRECTIONS

Together with the zeroth-order field construction, our main interest is in using symmetry to determine qualitative features such as whether particular mode propagation constants remain degenerate or whether there is a splitting when polarization effects, etc., are considered. Knowledge of these features provides a useful basis for quantification of eigenvalue and field corrections.

For formal quantification, we first note that manipulation of VWE and $\text{VWE}^{(0)}$ leads to an exact relation between their eigenvalues in terms of their fields [3]

$$\beta^2 - \tilde{\beta}^2 = \tilde{U}^2 - U^2/\rho^2 = <\tilde{\mathbf{e}}_t, \mathbb{H}_{pol}\, \mathbf{e}_t>/<\tilde{\mathbf{e}}_t, \mathbf{e}_t> \tag{2.21}$$

with

$$<\mathbf{a}, \mathbf{b}> \equiv \int_{A\infty} \mathbf{a}^* \bullet \mathbf{b}\, dA \quad \text{and} \quad \int_{A\infty} dA \equiv \int_{-\infty}^{\infty}\int_{-\infty}^{\infty} dx\,dy \equiv \int_{\phi=0}^{2\pi}\int_{r=0}^{\infty} r\,dr\,d\phi \tag{2.22}$$

with A_∞ being the infinite cross-section.

2.8.1 First-Order Eigenvalue Correction

As in the standard perturbation approach such as that applied to fibers by Snyder and Young [48] and Sammut et al. [50], given SWE (and $\text{VWE}^{(0)}$) eigenvalue $\tilde{U}$ together with $\text{VWE}^{(0)}$ field $\tilde{\mathbf{e}}_t$, expansion of Eq. (2.21) leads to first-order eigenvalue correction $U^{(1)}$ via

$$\begin{aligned} U^2 &= \tilde{U}^2 + \Delta < \tilde{\mathbf{e}}_t, \mathbb{H}_{p1}\, \tilde{\mathbf{e}}_t >/N + O(\Delta^2) \\ &= \tilde{U}^2 + 2\Delta\, \tilde{U}U^{(1)} + O(\Delta^2) \end{aligned} \tag{2.23}$$

where $$N = <\tilde{\mathbf{e}}_t, \tilde{\mathbf{e}}_t> \tag{2.24}$$

Note that if we write $\beta = \tilde{\beta} + \delta\beta$, to lowest order the polarization correction to the propagation constant is then given by

$$\delta\beta = -\frac{(2\Delta)^{3/2}\tilde{U}U^{(1)}}{2\rho V} \tag{2.25}$$

2.8.2 First-Order Field and Higher-Order Corrections

One method for obtaining the first-order field correction for mode j is to use the explicit relation [50]

$$\mathbf{e}_{tj}^{(1)} = \sum_i a_{ij}\tilde{\mathbf{e}}_{ti} \tag{2.26}$$

where the generalized summation is over all modes of the zeroth-order wave equation including an integration over the continuous spectrum of radiation modes, and where

$$a_{ij} = \begin{cases} H_{ij}\,(\dot{U}_j^2 - \dot{U}_i^2)N_j & \dot{U}_j \neq \dot{U}_i \quad \text{otherwise} \\ 0 \end{cases}$$

with $$H_{ij} \equiv <\tilde{\mathbf{e}}_{ti}, \mathbb{H}_{\mathbf{p1}}\,\tilde{\mathbf{e}}_{tj}> \tag{2.27}$$

given the standard degenerate perturbation theory assumption [91 (Sec. 10-3)] that $\tilde{\mathbf{e}}_{ti}$ have indeed been chosen as the linear combination of VWE$^{(0)}$ solutions which form the $\Delta \to 0$ limit of the exact fields. (Depending on symmetry-related simplifications etc., alternatives such as direct solution VWE$^{(1)}$ may be more efficient [50].) Given $\mathbf{e}_{tj}^{(1)}$ one may continue the expansion of Eq. (2.21) to obtain the second-order eigenvalue correction $U_j^{(2)}$ etc.

2.8.3 Simplifications due to Symmetry

As will be discussed in Sec. 3.4, symmetry can be exploited qualitatively to determine particular modal degeneracies without the necessity to explicitly evaluate the eigenvalue corrections, and quantitatively to simplify the evaluation of the diagonal terms H_{jj} = $<\tilde{\mathbf{e}}_{tj}, \mathbb{H}_{\mathbf{p1}}\tilde{\mathbf{e}}_{tj}>$ in the first-order eigenvalue corrections. For first-order corrections a symmetry approach is of particular tutorial value. For higher-order corrections which are more tedious to calculate, exploitation of symmetry can prove valuable in determination of which nondiagonal matrix elements H_{ij} contribute.

2.9 RECIPROCITY RELATIONS FOR ISOTROPIC MEDIA

It is often useful to relate the modal properties of two waveguides, e.g., (1) elliptical and circular fibers and (2) multicore and single-core fibers. The scalar mode propagation constant β corresponding to the modal field ψ of a guide with refractive index n is related to the propagation constant $\bar{\beta}$ of the mode $\bar{\psi}$ of a guide with index $\bar{n}$ by the reciprocity relation [3]

$$\beta^2 - \beta^2 = k^2 \int_{A\infty} (n^2 - \bar{n}^2)\psi\bar{\psi}\, dA \Big/ \int_{A_\infty} \psi\bar{\psi}\, dA \tag{2.28}$$

Approximating ψ in terms of $\bar{\psi}$ provides a basis for a perturbation approach to obtain β and modal properties in terms of those of a well-studied guide.

The vector version of the above reciprocity relation for nonabsorbing waveguides is [3 (Sec. 31-7), 92]

$$\beta^2 - \bar{\beta}^2 = k^2\sqrt{\varepsilon_0/\mu_0} \int_{A\infty} (n^2 - \bar{n}^2)\, \mathbf{e} \bullet \bar{\mathbf{e}}^*\, d\mathrm{A} \Big/ \int_{A\infty} \left\{\mathbf{e} \times \bar{\mathbf{h}}^* + \bar{\mathbf{e}}^* \times \mathbf{h}\right\} \bullet \hat{\mathbf{z}}\, dA \tag{2.29a}$$

$$\mathbf{h}_t = \sqrt{\varepsilon_0/\mu_0}\,\hat{\mathbf{z}} \times \left\{\beta \mathbf{e}_t + i\nabla_t e_z\right\}/k \qquad e_z = i\left\{\nabla_t \bullet \mathbf{e}_t + (\mathbf{e}_t \bullet \nabla_t)\ln n^2\right\}/\beta$$

$$(h_z = i\nabla_t \bullet \mathbf{h}_t/\beta) \tag{2.29b}$$

However, this is implicit in terms of β; we refer to Ref. 3 (Chap. 31) for alternative forms.

2.10 PHYSICAL PROPERTIES OF WAVEGUIDE MODES

In this book we concentrate on the determination of modal fields and propagation constants. From these quantities other basic modal properties are directly determined. We refer to Chap. 11 of Snyder and Love [3] for a full discussion. In this subsection we briefly summarize expressions for quantities of particular interest.

Given the modal propagation constant, the modal *phase* and *group* velocities are respectively given in terms of the parameters of Table 2.1*b* by

$$v_p = \frac{\omega}{\beta} = \frac{c}{n_{\text{eff}}} \quad \text{and} \quad v_g = \frac{d\omega}{d\beta} = -\frac{2\pi c}{\lambda^2}\frac{d\lambda}{d\beta} = \frac{c}{n_g} \tag{2.30}$$

where the modal effective (phase) index and group index are given by

$$n_{\text{eff}} = \frac{\beta}{k} \quad \text{and} \quad n_g = n_{\text{eff}} - \lambda\frac{dn_{\text{eff}}}{d\lambda} \tag{2.31}$$

In general, the total field propagating along a waveguide is given in terms of an expansion over the complete set of forward- and backward-propagating bound and radiation modes of a longitudinally invariant waveguide, each of the form given in Eq. (2.6), but with subscript j added to distinguish the different modes:

$$\begin{bmatrix}\mathbf{E}(\mathbf{r},z)\\ \mathbf{H}(\mathbf{r},z)\end{bmatrix} = \sum_j a_j \begin{bmatrix}\mathbf{E}_j(\mathbf{r},z)\\ \mathbf{H}_j(\mathbf{r},z)\end{bmatrix} = \sum_j a_j \begin{bmatrix}\mathbf{e}_j(\mathbf{r},z)\\ \mathbf{h}_j(\mathbf{r},z)\end{bmatrix} \exp(i\beta_j z) \tag{2.32}$$

where, as well as the discrete set of bound modes, the summation implicitly includes an integration over the continuum of radiation modes [3 (Chap. 25)].

Under the assumption of a nonabsorbing waveguide, the total power P carried by a mode and the fraction of modal power in the waveguide core η are given by

$$P = \frac{|a|^2}{2}\int_{A_\infty} \mathbf{e}\times\mathbf{h}^* dA \quad \text{and} \quad \eta = \int_{A_{\text{co}}} \mathbf{e}\times\mathbf{h}^* dA \Big/ \int_{A_\infty} \mathbf{e}\times\mathbf{h}^* dA \tag{2.33}$$

where the modal subscript j is implicit and A_{co} is the core cross section.

CHAPTER 3

Circular Isotropic Longitudinally Invariant Fibers

Even this seemingly innocent building block, the circular isotropic fiber, reveals a rich and instructive set of phenomena. Analysis may be in terms of several different mode sets: fixed or rotating scalar modes, vector modes that are true or pseudo with transverse or hybrid, linear or circular polarization. The appropriate wave equations, associated mode forms, and eigenvalues have interesting symmetry properties.

In Sec. 3.1 we briefly summarize the mode forms. Then, in Secs. 3.2 and 3.3, we provide a derivation and discussion of the mode forms and degeneracies in terms of symmetry. In particular, in Sec. 3.2, we introduce group theoretical methods for fibers using the circular fiber scalar modes as an illustration. Then, in Sec. 3.3, we consider a group theoretical framework for the construction of the true vector modes of weakly guiding [48] fibers. Finally, in Sec. 3.4, we provide quantitative evaluation of polarization eigenvalue level splitting.

3.1 SUMMARY OF MODAL REPRESENTATIONS

For azimuthally invariant index $n = n(r)$, we consider the transverse dependence for the field solutions of the wave equations of

Secs. 2.5 to 2.7. In particular, our major concern is with the scalar field $\psi(\mathbf{r})$ and, to zeroth order in Δ, the transverse component of the vector field $\tilde{\mathbf{e}}_t(\mathbf{r})$.

3.1.1 Scalar and Pseudo-Vector Mode Sets

In Table 3.1, we give the two sets of scalar mode field solutions of the scalar wave equation (SWE), Eq. (2.18), together with two associated sets of solutions of the zeroth-order vector wave equation, Eq. (2.17*a*), denoted by $\text{VWE}^{(0)}$.

3.1.2 True Weak-Guidance Vector Mode Set Constructions Using Pseudo-Modes

The latter vector modes form pseudo-mode sets. In general, particular combinations of LP [46] or CP [1] modes are required to obtain the weak-guidance limit of the true vector mode fields. Details of construction of both standard [3] and alternative [1, 70] true vector mode sets are given in Table 3.2. In particular, we see that to obtain $\tilde{\mathbf{e}}_t$ as a solution of $\text{VWE}^{(0)}$, which is also the zeroth-order term in the expansion (2.16*b*) of the exact VWE solution $\hat{\mathbf{e}}_t$, (1) the standard set requires a linear combination of two LP modes for azimuthal mode number $l > 0$, and (2) the alternative set requires two CP modes for construction of the transverse modes TM and TE, the only modes common to both sets. Note the alternative vector set replaces the hybrid HE or EH modes of the standard set by single CP modes.

3.1.3 Pictorial Representation and Notation Details

In Figs. 3.1 and 3.2, respectively, we give schematic representations of the fields of example modes from the standard and alternative sets.

Schematic Nature of Pictograms We emphasize that the mode forms are schematic only. For the vector mode *pictograms*, the arrows represent the electric field direction at the arrow centers. For the true vector mode pictograms given in the right column of Fig. 3.1, the arrows are placed at equispaced angular intervals on a circle of constant radius, and we have included enough arrows to

TABLE 3.1

Standard and Alternative Sets of Scalar and Zeroth-Order Vector Wave Equation Modes

Standard "Fixed" Mode Sets					Rotating Mode Sets				
l	*s*	Standard Scalar Modes	*k*	Linearly Polarized (LP) Modes	*l*	*s*	Rotating Scalar Modes	*k*	Circularly Polarized (CP) Vector Modes
$l = 0$	1	$\psi_{0m} = F_{0m}(R)$	1	$LP^{e}_{0m} = F_{0m}(R)\hat{\mathbf{x}}$	$l = 0$	1	$\psi_{0m} = F_{0m}(R)$	1	$CP^{R}_{0m} = F_{0m}(R)\hat{\mathbf{R}}$
			2	$LP^{e}_{0m} = F_{0m}(R)\,\hat{\mathbf{y}}$				2	$CP^{L}_{0m} = F_{0m}(R)\,\hat{\mathbf{L}}$
$l > 0$	1	$\psi^{e}_{lm} = F_{lm}(R)\cos l\phi$	1	$LP^{ex}_{lm} = F_{lm}(R)\,\hat{\mathbf{x}}\cos l\phi$	$l > 0$	1	$\psi^{+}_{lm} = F_{lm}(R)e^{il\phi}$	1	$CP^{R+}_{lm} = F_{lm}(R)\hat{\mathbf{R}}\,e^{il\phi}$
			2	$LP^{ey}_{lm} = F_{lm}(R)\,\hat{\mathbf{y}}\cos l\phi$				2	$CP^{L+}_{lm} = F_{lm}(R)\,\hat{\mathbf{L}}\,e^{il\phi}$
	2	$\psi^{o}_{lm} = F_{lm}(R)\sin l\phi$	1	$LP^{oy}_{lm} = F_{lm}(R)\,\hat{\mathbf{y}}\sin l\phi$		2	$\psi^{-}_{lm} = F_{lm}(R)e^{-il\phi}$	1	$CP^{R-}_{lm} = F_{lm}(R)\hat{\mathbf{R}}\,e^{-il\phi}$
			2	$LP^{ox}_{lm} = F_{lm}(R)\,\hat{\mathbf{x}}\sin l\phi$				2	$CP^{L-}_{lm} = F_{lm}(R)\,\hat{\mathbf{L}}\,e^{-il\phi}$

Notation: $F_{lm}(R)$ = radial field dependence = solution of $F'' + F'/R + (\bar{U}^2 - V^2 f - l^2/R^2)F = 0$, where dash indicates derivative with respect to $R \equiv r/\rho$; l = azimuthal mode number; m = radial mode number (irrelevant to the discussion of circular symmetry in Sec. 3.2 and thus sometimes omitted); mode labels s and k refer to Eq. (3.18) for the symmetry construction of true vector modes; ψ^{e}/ψ^{o} denotes that the ***scalar field distribution*** is even/odd with respect to the xz plane (Fig. 3.1); $\hat{\mathbf{R}} \equiv \{\hat{\mathbf{x}} - i\hat{\mathbf{y}}\}/\sqrt{2}$; $\hat{\mathbf{L}} \equiv \{\hat{\mathbf{x}} + i\hat{\mathbf{y}}\}/\sqrt{2}$; physical interpretation of +/– and R/L is given in Fig. 3.2.

TABLE 3.2

Standard and Alternative Sets of True Weak-Guidance Vector Modes and Their Transverse Field Polarizations

				Standard Hybrid True Mode Set				Alternative True Mode Set: Circularly Polarized Basis		
l	ν	h	p	Mode	Transverse Field Construction in Terms of LP Modes	Transverse Field $\mathbf{e}_{tlmp} = \mathbf{e}_{tlm}^{(\nu h)} = F_{lm}(R)\hat{\mathbf{p}}_h^{\nu l}(\phi)$ with Polarization $\hat{\mathbf{p}}_h^{\nu l}(\phi)$ in Linearly Polarized Basis	Polarization $\hat{\mathbf{p}}_h^{\nu l}(\phi)$ in Radially Polarized Basis	Mode	Transverse Field Relation of Circularly Polarized Set to Hybrid Mode Set	Transverse Field, $\mathbf{e}_{tlm}^{(\nu h)} = F_{lm}(R)\hat{\mathbf{p}}_h^{\nu l}(\phi)$ with Polarization in Circularly Polarized Basis
0	**1**	1	1	HE_{1m}^{e}	LP_{0m}^{e}	$F_{0m}(R)\hat{\mathbf{x}}$	$\hat{\mathbf{r}}\cos\phi - \hat{\phi}\sin\phi$	CP_{0m}^{R}	$\frac{1}{\sqrt{2}}\{HE_{1m}^{e} - iHE_{1m}^{\theta}\}$	$F_{0m}(R)\hat{\mathbf{R}}$
		2	3	HE_{1m}^{o}	LP_{0m}^{e}	$F_{0m}(R)\hat{\mathbf{y}}$	$\hat{\mathbf{r}}\sin\phi + \hat{\phi}\cos\phi$	CP_{0m}^{L}	$\frac{1}{\sqrt{2}}\{HE_{1m}^{e} + iHE_{1m}^{0}\}$	$F_{0m}(R)\hat{\mathbf{L}}$
1	**0**	1	2	TM_{0m}	$LP_{1m}^{ex} + LP_{1m}^{oy}$	$F_{1m}(R)\{\hat{\mathbf{x}}\cos\phi + \hat{\mathbf{y}}\sin\phi\}$	$\hat{\mathbf{r}}$	TM_{0m}	$\frac{1}{\sqrt{2}}\{CP_{1m}^{R+} + CP_{1m}^{L-}\}$	$F_{1m}(R)\frac{\hat{\mathbf{R}}e^{i\phi} + \hat{\mathbf{L}}e^{-i\phi}}{\sqrt{2}}$
	$\tilde{\mathbf{0}}$	1	4	TE_{0m}	$LP_{1m}^{ox} - LP_{1m}^{ey}$	$F_{1m}(R)\{\hat{\mathbf{x}}\sin\phi - \hat{\mathbf{y}}\cos\phi\}$	$\hat{\phi}$	TE_{0m}	$\frac{-i}{\sqrt{2}}\{CP_{1m}^{R+} - CP_{1m}^{L-}\}$	$F_{1m}(R)\frac{\hat{\mathbf{R}}e^{i\phi} + \hat{\mathbf{L}}e^{-i\phi}}{\sqrt{2}}$
	2	1	1	HE_{2m}^{e}	$LP_{1m}^{ex} - LP_{1m}^{oy}$	$F_{1m}(R)\{\hat{\mathbf{x}}\cos\phi - \hat{\mathbf{y}}\sin\phi\}$	$\hat{\mathbf{r}}\cos 2\phi - \hat{\phi}\sin 2\phi$	CP_{1m}^{R-}	$\frac{1}{\sqrt{2}}\{HE_{2m}^{e} - iHE_{2m}^{o}\}$	$F_{1m}(R)\hat{\mathbf{R}}e^{-i\phi}$
		2	3	HE_{2m}^{o}	$LP_{1m}^{ox} + LP_{1m}^{ey}$	$F_{1m}(R)\{\hat{\mathbf{x}}\sin\phi + \hat{\mathbf{y}}\cos\phi\}$	$\hat{\mathbf{r}}\sin 2\phi + \hat{\phi}\cos 2\phi$	CP_{1m}^{L+}	$\frac{1}{\sqrt{2}}\{HE_{2m}^{e} + iHE_{2m}^{o}\}$	$F_{1m}(R)\hat{\mathbf{L}}e^{i\phi}$

$l > 1$	$l-1$	1	2	$\mathrm{EH}^{e}_{l-1,m}$	$\mathrm{LP}^{ex}_{lm} + \mathrm{LP}^{oy}_{lm}$	$F_{lm}(R)\{\hat{\mathbf{x}}\cos l\phi + \hat{\mathbf{y}}\sin\phi\}$	$\hat{\mathbf{r}}\cos(l-1)\phi + \hat{\phi}\sin(l-1)\phi$	CP^{L-}_{lm}	$\frac{1}{\sqrt{2}}\{\mathrm{EH}^{e}_{l-1,m} + i\mathrm{EH}^{o}_{l-1,m}\}$	$F_{lm}(R)\hat{\mathbf{L}}e^{-il\phi}$
		2	4	$\mathrm{EH}^{o}_{l-1,m}$	$\mathrm{LP}^{ox}_{lm} - \mathrm{LP}^{ey}_{lm}$	$F_{lm}(R)\{\hat{\mathbf{x}}\sin l\phi - \hat{\mathbf{y}}\cos l\phi\}$	$\hat{\mathbf{r}}\sin(l-1)\phi - \hat{\phi}\cos(l-1)\phi$	CP^{R+}_{lm}	$\frac{1}{\sqrt{2}}\{\mathrm{EH}^{e}_{l-1,m} - i\mathrm{EH}^{o}_{l-1,m}\}$	$F_{lm}(R)\hat{\mathbf{R}}e^{il\phi}$
	$l+1$	1	1	$\mathrm{HE}^{e}_{l+1,m}$	$\mathrm{LP}^{ex}_{lm} - \mathrm{LP}^{oy}_{lm}$	$F_{lm}(R)\{\hat{\mathbf{x}}\cos l\phi - \hat{\mathbf{y}}\sin l\phi\}$	$\hat{\mathbf{r}}\cos(l+1)\phi - \hat{\phi}\sin(l+1)\phi$	CP^{R-}_{lm}	$\frac{1}{\sqrt{2}}\{\mathrm{HE}^{e}_{l+1,m} - i\mathrm{HE}^{o}_{l+1,m}\}$	$F_{lm}(R)\hat{\mathbf{R}}e^{-il\phi}$
		2	3	$\mathrm{HE}^{o}_{l+1,m}$	$\mathrm{LP}^{ox}_{lm} + \mathrm{LP}^{ey}_{lm}$	$F_{lm}(R)\{\hat{\mathbf{x}}\sin l\phi + \hat{\mathbf{y}}\cos l\phi\}$	$\hat{\mathbf{r}}\sin(l+1)\phi + \hat{\phi}\cos(l+1)\phi$	CP^{L+}_{lm}	$\frac{1}{\sqrt{2}}\{\mathrm{HE}^{e}_{l+1,m} + i\mathrm{HE}^{o}_{l+1,m}\}$	$F_{lm}(R)\hat{\mathbf{L}}e^{il\phi}$

Notation: For azimuthal mode number l, the natural symmetry classification of the different polarizations is νh. Alternatively the single polarization mode number p may be used corresponding to the numbering in Ref. 3. As distinct from the *scalar* mode *e/o* (even/odd) nomenclature for ψ^e/ψ^o in Table 3.1 denoting that the *scalar field distribution* is even/odd with respect to the xz plane, the *polarization e/o* nomenclature $\mathrm{HE}^e/\mathrm{HE}^o$ and $\mathrm{EH}^e/\mathrm{EH}^o$ corresponds to the *z-polarized component* (and thus x component) being even/odd with respect to the xz plane. That is, for azimuthal mode number l, with $\nu = l \pm 1$, we have $e^{+e}_{zlm} = aG^{\mp}\cos(l \pm 1)\phi$ and $e^{\pm 0}_{zlm} = aG^{\mp}\sin(l \pm 1)\phi$, where $a = i\sqrt{2\Delta}/V$ and $G^{\pm}_{lm} = F' \pm lF/R$.

FIGURE 3.1

Lowest-order modes of a single-core circularly symmetric fiber. If the azimuthal mode number $l = 0$, then the true mode is approximately LP; for $l > 0$, each true mode (within the weak-guidance approximation) is constructed from an equal combination of two LP modes. For singly-clad fibers N provides an approximate ordering of the modal eigenvalues. If the radial profile is parabolic, then modes with the same N (for example, lm = 21 and 02) are degenerate. A right-handed coordinate system with propagation along positive $\mathbf{z}$ into the page is used throughout. **Even/odd** (*e/o*) is with respect to the xy plane: for scalar and LP modes it denotes field magnitude distribution; for true modes it denotes the distribution of $\mathbf{e}_x$ (or equivalently $\mathbf{e}_z$). Mode forms are schematic only. **Arrows** represent **at their centers** the transverse electric field direction (and approximate relative magnitude). They are centered at equispaced intervals in the azimuthal coordinate ϕ. Note that $\pm HE_{\nu m}/EH_{\nu m}$ odd mode patterns may be obtained by $(2n + 1)$ π-$(2n)$ rotations (n = 0, 1, . . .) of the corresponding even modes.

lm	$N = l + 2(m-1)$	Standard scalar modes		LP pseudo = vector modes		Standard true vector modes	
01	0	ψ_{01}		LP_{01}^x	LP_{01}^y	HE_{11}^e	HE_{11}^o
11	1	ψ_{11}^e	ψ_{11}^o	LP_{11}^{ex} LP_{11}^{ox}	LP_{11}^{oy} LP_{11}^{ey}	TE_{01} $(\approx LP_{11}^{ox} - LP_{11}^{ey})$; TM_{01} $(\approx LP_{11}^{ex} + LP_{11}^{oy})$; HE_{21}^e $(\approx LP_{11}^{ex} - LP_{11}^{oy})$	HE_{21}^o $(\approx LP_{11}^{ox} + LP_{11}^{ey})$
21	2	ψ_{21}^e	ψ_{21}^o	LP_{21}^{ex} LP_{21}^{ox}	LP_{21}^{ey} LP_{21}^{oy}	EH_{11}^e $(\approx LP_{21}^{ex} + LP_{21}^{oy})$; HE_{31}^e $(\approx LP_{21}^{ex} - LP_{21}^{oy})$	EH_{11}^o $(\approx LP_{21}^{ox} - LP_{21}^{ey})$; HE_{31}^o $(\approx LP_{21}^{ox} + LP_{21}^{ey})$
02	2	ψ_{02}		LP_{02}^x	LP_{02}^y	HE_{12}^e	HE_{12}^o

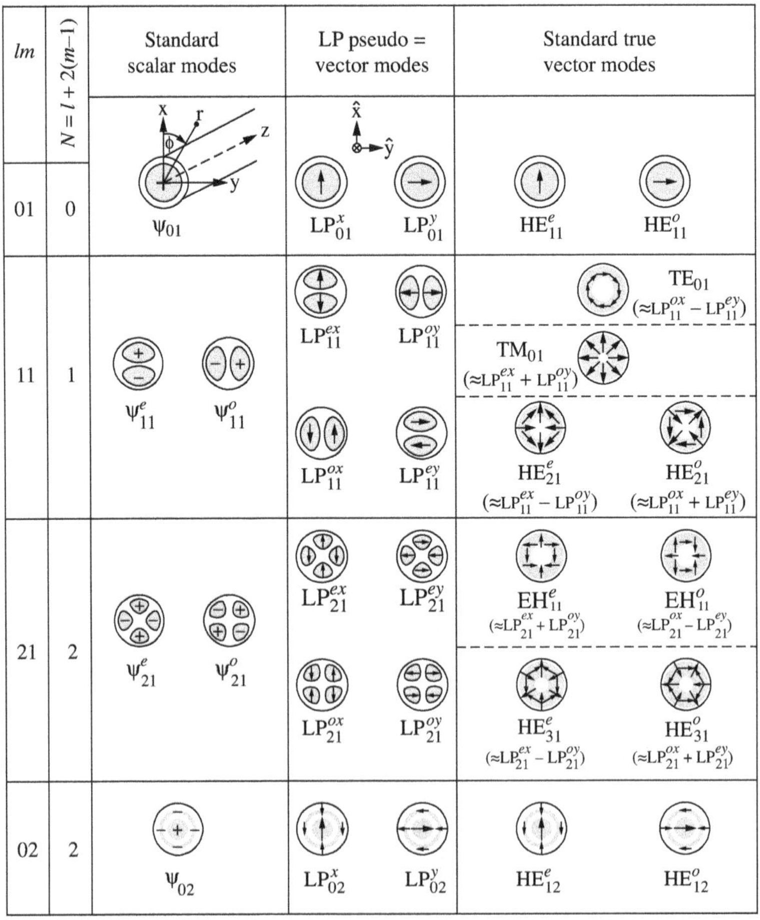

FIGURE 3.2

Circularly polarized modes. (*a*) *Scalar modes*: +/– corresponds to a left/right rotating scalar mode intensity pattern, i.e., clockwise/anticlockwise for an observer who, by convention, looks back at the oncoming wave that propagates in the +*z* direction (or the anticlockwise/clockwise if we look from behind the wave as in the figure). (*b*) *Vector modes*: *L*/*R* correspond to left/right circular polarization of the vector mode local field vectors, i.e., by convention clockwise/anticlockwise for an observer looking at the oncoming wave.

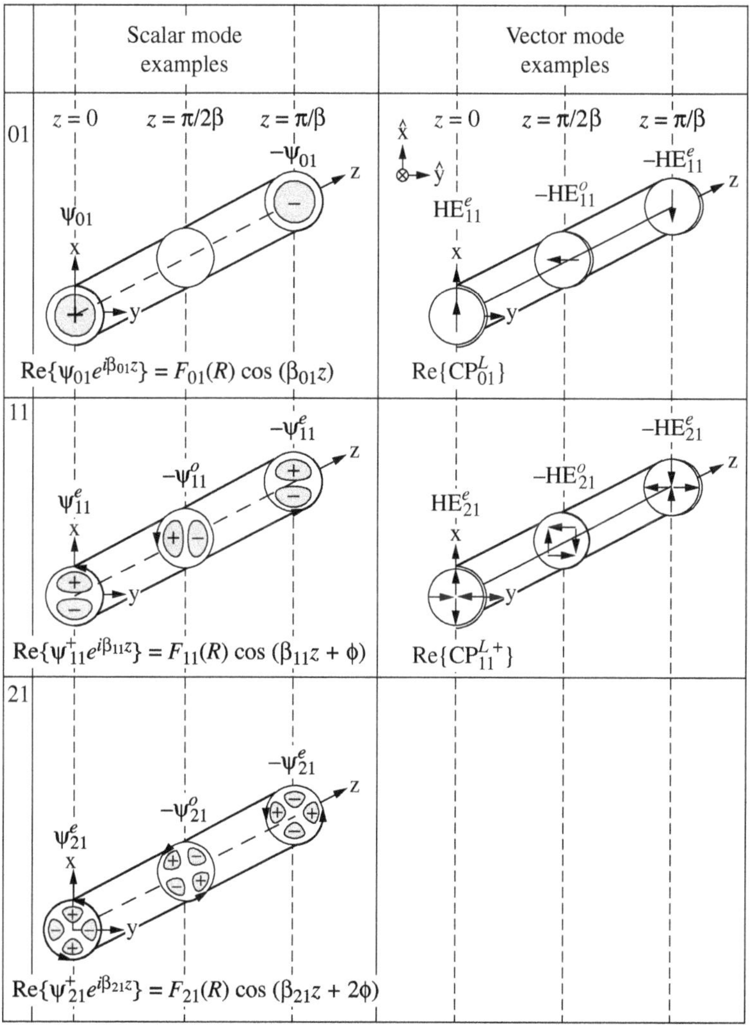

show that rotation of a hybrid mode pattern by $\pi/2\nu$ will convert an even to an odd mode. Thus, for HE_{21} the pictograms have the field direction given at angular intervals of $\pi/4$, for HE_{31} at intervals of $\pi/6$.

Simplified Pictograms Often, it is convenient to use simplified pictograms with the minimum number of arrows to distinguish each mode from the others, i.e., at intervals of $\pi/2l$ in the azimuthal coordinate. In particular, for the second true mode set, we often use the forms with the field direction given only at azimuthal intervals of $\pi/2$:

For the third mode set, i.e., the EH_{11}/HE_{31} mode pairs, $\pi/4$ intervals are sufficient:

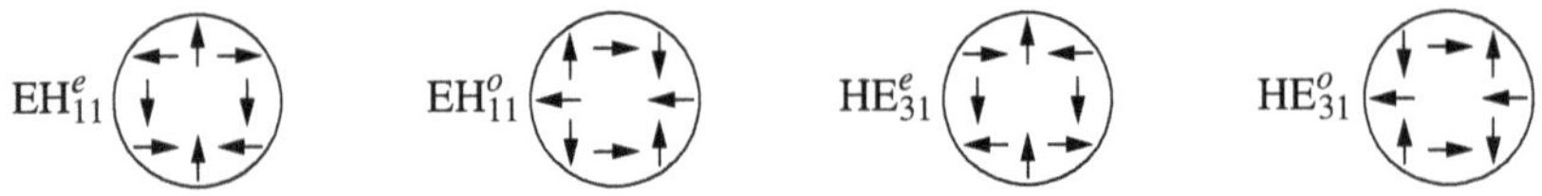

3.2 SYMMETRY CONCEPTS FOR CIRCULAR FIBERS: SCALAR MODE FIELDS AND DEGENERACIES

In this section, we introduce the application of symmetry concepts to fibers by using the simplest example: circular fiber scalar modes. This exploits the matrix representation theory of the group of symmetry operations corresponding to a circle, which is discussed in the Appendix. Although the scalar case is fairly trivial and well known particularly for the case of the identical two-dimensional Schrödinger equation, its treatment will aid our understanding of symmetry concepts required for the novel construction of vector modes discussed in Sec. 3.3.

§ Although analogies do exist [1, 93], the details of the vector problem are different from those found in quantum mechanics. §

Furthermore, as both these sections will provide a basis for the rest of the chapter where we give results for more complex geometries and anisotropic materials, we discuss the methods, symmetry operations, etc. in considerable detail. However, we remark that once the appropriate wave equation symmetries are recognized, one can immediately write down the essential results for the symmetry determination of the SWE, $VWE^{(0)}$, and VWE modal field constructions and degeneracies that are summarized in Sec. 3.2.3, Table 3.3, and especially Table 3.4.

Essentially the results that we develop in these two sections may be summarized as follows. The azimuthal dependencies of circular fiber modal fields are entirely determined as basis functions corresponding to matrix representations of the symmetry operations associated with the appropriate wave equation symmetry group. Different matrix representations correspond to different mode propagation constant levels, and thus association of modes with these representations allows determination of propagation constant degeneracies and splittings.

In this section, after discussing the geometrical and wave equation symmetries, in Sec. 3.2.3 we summarize the consequences of symmetry for the SWE mode forms and degeneracies, and then in Sec. 3.2.4 we provide an intuitive understanding of why these are required by symmetry, referring to the Appendix for formal group theoretic details.

TABLE 3.3

Symmetry Reduction: SWE ⊗ Polarization → Joint and Associated Pseudo-Modes and True Modes of a Circular Core Fiber in Terms of Standard Direct Product Irrep Branching Rules for $\mathbf{C}_{\infty v} \otimes \mathbf{C}_{\infty v} \supset \mathbf{C}_{\infty v}$

Symmetry Groups	$\mathbf{C}^{S}_{\infty v} \otimes \mathbf{C}^{P}_{\infty v} \supset \mathbf{C}^{J}_{\infty v}$	Degeneracies and Modes
Branching Rule	$l \otimes 1 \rightarrow \sum_{v} n^{v}_{l l} v$	$\|l \otimes 1\|$ $\text{pseudo}_{lm} \rightarrow$ $\|v\|$ true_{vm} degeneracies modes
$l = 0$	$\mathbf{0 \otimes 1 \rightarrow 1}$	$2LP_{0m} \rightarrow 2HE_{1m}$
$l = 1$	$\mathbf{1 \otimes 1 \rightarrow 0 \oplus \tilde{0} \oplus 2}$	$4LP_{1m} \rightarrow 1TM_{0m} + 1TE_{0m} + 2HE_{2m}$
$l > 1$	$\mathbf{l \otimes 1 \rightarrow (l-1) \oplus (l+1)}$	$4LP_{lm} \rightarrow 2EH_{l-1,m} + 2HE_{l+1,m}$

The direct-product branching rules (also known as Kronecker product rules or Clebsch-Gordan series) given in the second column may be obtained from Table 10, p. 17 of Atkins et al. [10] or straightforwardly by using group character theory, which gives the coefficients n^{v}_{l1} as described in Sec. A.3 (Example 2). In applications here we will often add indices to the irreps, for example, $0^{S}_{\infty v} \otimes 1^{P}_{\infty v} \rightarrow 1^{J}_{\infty v}$ as an abbreviation for $0(\mathbf{C}^{S}_{\infty v}) \otimes 1(\mathbf{C}^{P}_{\infty v}) \rightarrow 1(\mathbf{C}^{J}_{\infty v})$ to emphasize the groups corresponding to SWE, polarization and joint symmetries, respectively (always this order). Also note that (1) labels in boldface parentheses, e.g., $(\mathbf{l-1})$, correspond to single irreps, and (2) vertical lines as in $|v|$ indicate irrep dimensions.

TABLE 3.4

Circular Isotropic Fiber Wave Equation Symmetries and Associated Modal Constructions in Terms of Irrep Basis Functions.

Wave Equation Symmetry	**Symmetry Group** operation	**Irreps** (**b** = **basis** **r** = **real;** **c** = **complex helical)**		ϕ–**Dependent Basis Functions** [To obtain (R,ϕ)-dependent basis functions, these may be multiplied by a general function $F(R)$ as R is invariant under $\mathbf{C}_{\infty v}$] ($\otimes$ signifies all possible products)	**Basis Function Nomenclature** {BasFunc$_l(\phi)$}	**Modal Fields** ψ or $\tilde{\mathbf{e}}_t = F_{lm}(R)$ BasFunc$_l(\phi)$	**Mode** (Tables 3.1 and 3.2)
(*a*) **SWE**	$\mathbf{C}^{\mathrm{S}}_{\infty v}$	$l = 0$	–**r, c**	$\{1\}$	$\{\Phi_1^1\}$	$\psi_{0m} = F_{0m}(R)$	ψ_{0m}
$g_{\mathrm{S}}\psi(\mathbf{r}) = \psi(g^{-1}\mathbf{r})$		$l \geq 1$	–**r**	$\{\cos l\phi, \sin l\phi\}$	$\{\Phi_s^l(\phi) : s = 1, 2\}$	$\psi_{lm}^s = F_{lm}(R)\Phi_s^l(\phi)$	ψ_{lm}^s
			–**c**	$\{e^{il\phi}, e^{-il\phi}\}$			
(*b*) **VWE**$^{(0)}$	$\mathbf{C}^{\mathrm{S}}_{\infty v} \otimes \mathbf{C}^{\mathrm{J}}_{\infty v}$	$\mathbf{0} \otimes \mathbf{1}$, $l = 0$	–**r**	$\{1\} \otimes \{\hat{\mathbf{x}}, \hat{\mathbf{y}}\} \equiv \{\hat{\mathbf{x}}, \hat{\mathbf{y}}\}$	$\{\hat{\mathbf{p}}_k : k = 1, 2\}$	$\tilde{\mathbf{e}}_t = F_{0m}(R)\hat{\mathbf{p}}_k$	LP_{0m}^k
			–**c**	$\{1\} \otimes \{\hat{\mathbf{R}}, \hat{\mathbf{L}}\} \equiv \{\hat{\mathbf{R}}, \hat{\mathbf{L}}\}$		$= \psi_{0m}(R)\hat{\mathbf{p}}_k$	CP_{0m}^k
$g_{\mathrm{S}}h_{\mathrm{P}}\tilde{\mathbf{e}}_t(\mathbf{r}) = \sum_{k=1}^{2}\{g_{\mathrm{S}}\tilde{\mathbf{e}}_k(\mathbf{r})\{h_p\hat{\mathbf{p}}_k\}$		$\boldsymbol{l} \otimes \mathbf{1}$, $l \geq 1$	–**r**	$\{\cos l\phi, \sin l\phi\} \otimes \{\hat{\mathbf{x}}, \hat{\mathbf{y}}\}$	$\{\hat{\mathbf{p}}_{sk}^l(\phi) \equiv \Phi_s^l(\phi)\hat{\mathbf{p}}_k :$	$\hat{\mathbf{e}}_t = F_{lm}(R)\mathbf{p}_{sk}^l(\phi)$	LP_{lm}^{sk}
$g_{\mathrm{S}} \in \mathrm{C}^{\mathrm{S}}_{\infty v}$, $h_p \in \mathrm{C}^{\mathrm{P}}_{\infty v}$				$\equiv \cos l\phi\ \hat{\mathbf{x}}, \cos l\phi\ \hat{\mathbf{y}}, \sin l\phi\ \hat{\mathbf{x}}, \sin l\phi\ \hat{\mathbf{y}}\}$	$s = 1, 2; k = 1, 2\}$	$= F_{lm}(R)\Phi_s^l(\phi)\hat{\mathbf{p}}_k$	
			–**c**	$\{e^{il\phi}, e^{il\phi}\} \otimes \{\hat{\mathbf{R}}, \hat{\mathbf{L}}\}$		$= \psi_{lm}^s(R,\phi)\hat{\mathbf{p}}_k$	CP_{lm}^{sk}

(*c*) VWE $\mathbf{C}^{S}_{\infty v} \otimes \mathbf{C}^{P}_{\infty v} \subset \mathbf{C}^{J}_{\infty v}$	$\nu = 0$	$-\mathbf{r}, \mathbf{c}$	$\{\hat{r} = (\hat{\mathbf{R}} e^{i\phi} + \mathbf{L} e^{-i\phi})/\sqrt{2}\}$	$\{\hat{\mathbf{p}}_1^{\bar{0}1}(\phi)\}$	$\tilde{\mathbf{e}}_t = F_{1m}(R)\hat{\mathbf{r}}(\phi)$	TM_{0m}
	$\nu = \tilde{0}$	$-\mathbf{r}, \mathbf{c}$	$\{\hat{\phi} = (\hat{\mathbf{R}} e^{i\phi} - \mathbf{L} e^{-i\phi})/\sqrt{2}\}$	$\{\hat{\mathbf{p}}_1^{01}(\phi)\}$	$\tilde{\mathbf{e}}_t = F_{1m}(R)\hat{\phi}(\phi)$	TE_{0m}
$g_J \tilde{\mathbf{e}}_t(\mathbf{r}) = \sum_{k=1}^{2} \{g_S \tilde{e}_k(\mathbf{r})\}\{g_P \hat{\mathbf{p}}_k\}$	$\nu \geq l$	$-\mathbf{r}, \mathbf{c}$	$\{(\sum_{s=1}^{2}\sum_{k=1}^{2} c_{hsk}^{\nu l} \Phi_s^l(\phi)\hat{\mathbf{p}}_k) : h = 1, 2\}_{b=r \text{ or } c}$	$\{\hat{\mathbf{p}}_h^{\nu l}(\phi) : h = 1, 2\}$	$\{\hat{\mathbf{p}}_h^{\nu l}(\phi) :\ h = 1,2\}$	<u>**b=r**</u> $\nu = l + 1 : \mathrm{HE}_{\nu m}^{h}$
i = J,S,P $g_i \equiv O_i(g) \in \mathbf{C}^{i}_{\infty \mathbf{v}}$, = joint,			$\Phi_s^l / \hat{\mathbf{p}}_k$—see SWE/VWE$^{(0)}$ basis functions			$\nu = l - 1 : \mathrm{EH}_{\nu m}^{h}$ <u>**b=c**</u>
scalar, $g \in \mathbf{C}_{\infty \mathbf{v}}$. polarization symmetry			$c_{hsk}^{\nu l} = \langle ls, 1k / \nu h \rangle_{b=r \text{ or } c}$ = Clebsch–Gordan coefficients (Sec. A.4)			$\nu = l \pm 1 : \mathrm{CP}_{lm}^{sk}$

(***d***) *Nomenclature*: $h = 1, \ldots, |\nu|$ = true mode polarization number
$s = 1, \ldots, |l|$ = scalar mode parity no., i.e., $s \in \{1, 2\} = \{e, p\}$ or $\{+, -\}$
$k = 1, 2$ = pseudo-mode polarization no. $k \in \{1, 2\} = (x, y)$ or $\{R, L\}$
$|\nu|$ = degeneracy of true mode level ν = dimension of irrep ν
$|l|$ = degeneracy of scalar mode level
l = dimension of irrep l
i.e., $|\nu| = 1$ for $\nu = 0$ or $\tilde{0}$ and $|\nu| = 2$ otherwise; $|l| = 1$ for $l = 0$ and $|l| = 2$ for $l > 0$

3.2.1 Geometrical Symmetry: $\mathbf{C}_{\infty v}$

For a circular isotropic fiber, the azimuthally invariant refractive index $n = n(r)$ is invariant under the symmetry operations of a circle: these correspond to rotations of any angle about the propagation axis as well as two-fold reflections in any plane passing through the propagation z axis; note that reflection in any such plane can be obtained by a rotation followed by a reflection in the xz plane. The appropriate *symmetry group corresponding to the circular fiber structure* that includes these symmetry operations is the *two-dimensional rotation-reflection group* $\mathbf{C}_{\infty v}$, which is sometimes referred to as **O(2)** or $\mathbf{O}_2$ and whose properties are described in Sec. A.2 (Tables A.2 to A.4).

§ In later sections we will consider refractive indices $n = n(r, \phi)$ for which the rotations must be by discrete angles $2\pi/n$ to leave the structure invariant, and thus the symmetry will be reduced to the discrete rotation-reflection group $\mathbf{C}_{nv}$. §

3.2.2 Scalar Wave Equation Symmetry: $\mathbf{C}^{S}_{\infty v}$

It may be shown that the two-dimensional laplacian is also invariant under $\mathbf{C}_{\infty v}$, and hence for a general index profile the symmetry of the scalar wave equation (SWE) operator $\mathcal{H}_s = \nabla_t^2 + k^2 n^2(r, \phi)$ reverts to consideration of the azimuthal dependence of the index profile [1]. Thus for the general circular isotropic fibers considered in this section, the SWE is also invariant under $\mathbf{C}_{\infty v}$.

§ Hidden Symmetry We remark that certain special radial dependencies of $n(r)$ lead to an enlargement of the symmetry group of the equation as a whole, i.e., *hidden* or *dynamical* symmetry resulting in extra modal degeneracies, which we consider for fibers in [1]. For example, for parabolic profiles, $\mathcal{H}_S$ has additional symmetry related to separability as $\mathcal{H}(x) + \mathcal{H}(y)$ as well as the usual circular profile separability $\mathcal{H}(r) + \mathcal{H}(\phi)$; this is described by the matrix group $\mathbf{SU}_2$ (Sec. A.1 and Ref. 6, Chap. 19). §

For the purpose of distinguishing the symmetry of $\mathcal{H}_S$ from other symmetries introduced later, it is convenient to label the associated group with superscript S, that is, $\mathbf{C}^{S}_{\infty v}$ for circular fibers.

General SWE operator symmetry group: In general, we **define $\mathbf{G}^S$ to be the group of the SWE operator $\mathcal{H}_S$** and **to consist of operators $g_S \equiv O_S(g)$, which act specifically on scalar functions $\psi(\mathbf{r})$**; that is, $O(g)\psi(\mathbf{r}) \equiv g_S\psi(\mathbf{r})$ in the induced transformation of Eq. (A.1), which defines the relation of $\mathbf{G}^S$ to an associated well-known group $\mathbf{G}$.

Furthermore, if the associated group $\mathbf{G}$ is a group of coordinate transformations g acting on a position vector $\mathbf{r}$, that is, $O(g)\mathbf{r} \equiv g\mathbf{r}$ (as is our view of $\mathbf{C}_{\infty v}$), then Eq. (A.1) takes the form

$$g_S\psi(\mathbf{r}) = \psi(g^{-1}\mathbf{r})$$

3.2.3 Scalar Modes: Basis Functions of Irreps of $\mathbf{C}^S_{\infty v}$

The immediate formal consequences of circular $\mathbf{C}_{\infty v}$ symmetry (for which we will provide a detailed explanation in the next subsection) may be summarized as follows:

1. The SWE has modal solutions of separable form

$$\psi^S_{lm}(R,\phi) = F_{lm}(R)\Phi^l_S(\phi) \qquad (3.1a)$$

where

(*a*) The angular dependence $\Phi^l_S(\phi)$ is given by basis functions for the irreducible matrix representations l of $\mathbf{C}_{\infty v}$ (sometimes written $\mathbf{D}^{(l)}$ and given in Table A.4), that is, using the mathematical symbol $\in$ denoting "is a member of the set," we have

$$\Phi^l_S \in \{\cos l\phi, \sin l\phi\} \quad \text{or} \quad \Phi^l_S \in \{e^{il\phi}, e^{-il\phi}\} \qquad (3.1b)$$

which are, respectively, real and complex exponential representation scalar basis functions (SBFs in Table A.4) giving the forms of standard and rotating scalar mode sets in Table 3.1.

and

(*b*) $\mathbf{C}_{\infty v}$ symmetry places no restriction on F apart from its being ϕ-independent. Its form is determined by substitution into the SWE, which leads to the l-dependent radial wave equation given in the caption of Table 3.1. Then $F_{lm}(R)$ is the mth solution of this equation.

2. The *degeneracies* of the SWE eigenvalues $\widetilde{U}_{lm}$ (or propagation constants $\widetilde{\beta}_{lm}$) are given by the *dimensions* of the irreps $l(\mathbf{C}_{\infty v})$, that is, 1 for $l = 0$ and 2 for $l \geq 1$.

3.2.4 Symmetry Tutorial: Scalar Mode Transformations

To aid our intuitive understanding of the consequences of circular symmetry for the mode forms, first recall that a mode is any field entity that propagates a particular phase velocity: thus any linear combination of modes with the same propagation constant will also be a mode. Given this definition, application of a symmetry operation of a circle to a mode will also lead to a mode; this may be the original mode, another mode with the same propagation constant, or a linear combination of the two. If we apply the same symmetry operation to each mode of a set of orthogonal modes, then we obtain an alternative set.

In particular, if we examine the scalar modes given in Fig. 3.1 and consider the symmetry operations on the scalar field ψ of (1) reflection σ_{vS} in the xz plane and (2) rotations, $C_S(\theta)$ of angle θ about the z axis, corresponding to the usual coordinate transformations σ_v and C_θ, then

(*a*) The azimuthally independent $l = 0$ modes obviously remain invariant under both reflections and rotation, that is, $\sigma_{vS}\psi_{0m} = C_S(\theta)\,\psi_{0m} = \psi_{0m}$.

(*b*) For $l \geq 1$, *reflection* σ_{vS} leads to the original mode within a sign, that is

$$\sigma_{vS}[\psi^e, \psi^o] = [\psi^e, -\psi^o] = [\psi^e, \psi^o]\begin{bmatrix} 1 & 0 \\ 0 & -1 \end{bmatrix} \tag{3.2}$$

(*c*) Rotations by $\theta = \pi/2l$ convert an even mode to the corresponding odd mode and vice versa within a sign, that is, $C_S(\pi/2l)\psi^o_{lm} = \psi^e_{lm}$ and $C_S(\pi/2l)\psi^e_{lm} = -\psi^o_{lm}$. Thus, *rotation being a symmetry operation for the system means that* ψ^e_{lm} and ψ^o_{lm} *must have the same propagation constant.*

(*d*) More general rotations mix modes ψ^e and ψ^o, producing alternative ones ψ^{a1} and ψ^{a2}. In particular, we note the induced transformation $\psi^a(R, \phi) \equiv C_S(\theta)\psi(R, \phi) = \psi(R, \phi - \theta)$, which says that a rotated mode is functionally equivalent to an unrotated mode viewed from inversely rotated coordinates, together with the form of the latter *passive rotation* as

$$\begin{bmatrix} \psi^e_{lm}(R,\phi-\theta) \\ \psi^o_{lm}(R,\phi-\theta) \end{bmatrix} = F_{lm}(R)\begin{bmatrix} \cos l(\phi-\theta) \\ \sin l(\phi-\theta) \end{bmatrix}$$

$$= F_{lm}(R)\begin{bmatrix} \cos l\theta & \sin l\theta \\ -\sin l\theta & \cos l\theta \end{bmatrix}\begin{bmatrix} \cos l\phi \\ \sin l\phi \end{bmatrix} \tag{3.3a}$$

This leads to an alternative set of *actively rotated* modes given by

$$\begin{bmatrix} \psi^{a1}_{lm}(R,\phi) \\ \psi^{a2}_{lm}(R,\phi) \end{bmatrix} = C_S(\theta)\begin{bmatrix} \psi^e_{lm}(R,\phi) \\ \psi^o_{lm}(R,\phi) \end{bmatrix}$$

$$= \begin{bmatrix} \cos l\theta\psi^e_{lm}(R,\phi)+\sin l\theta\psi^o_{lm}(R,\phi) \\ -\sin l\theta\psi^e_{lm}(R,\phi)+\cos l\theta\psi^o_{lm}(R,\phi) \end{bmatrix} \tag{3.3b}$$

Conventionally, this is written in transposed form [assuming coordinates (R, ϕ) for ψ] as

$$C_S(\theta)\,[\psi^e_{lm}, \psi^o_{lm}] = [\psi^o_{lm}, \psi^o_{lm}]\begin{bmatrix} \cos l\theta & -\sin l\theta \\ \sin l\theta & -\cos l\theta \end{bmatrix} \tag{3.3c}$$

For example, pictorially, we have the rotation of ψ^e_{11} by $\theta = \pi/4$ as

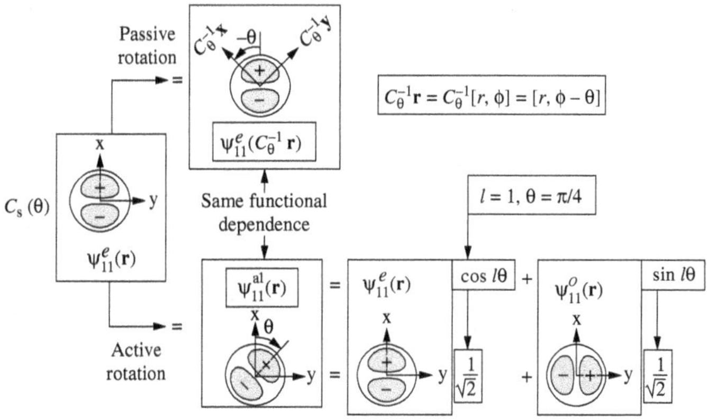

(3.4)

In terms of the group representation theory of App. A, and in particular the real irreps and basis functions of Table A.4, these transformation results are interpreted as follows.

1. The rotation-reflection invariance of ψ_{0m} corresponds to $\mathbb{D}^{(0)}(C_\theta) = \mathbb{D}^{(0)}(\sigma_v) = 1$. Thus the modes ψ_{0m} are said to transform as the one-dimensional matrix representation **0** (often referred to as $\mathbf{D}^{(0)}$), which corresponds to one ϕ-independent basis function.
2. For $l \geq 1$, we note that the rotation matrix in Eq. (3.3*c*) corresponds to $\mathbb{D}^{(l)}(C_\theta)$ in Table A.4 with the reflection matrix in Eq. (3.2) being $\mathbb{D}^{(l)}(\sigma_v)$. Then these are simply matrix forms of Eq. (A.3), which defines basis functions associated with a matrix representation. Thus the standard set of scalar modes ψ_{lm} is said to transform as the real two-dimensional matrix representations $\boldsymbol{l}$ ($\equiv \mathbf{D}^{(l)}$), which each have two basis functions with ϕ dependence $[\cos l\phi, \sin l\phi]$.

3.3 VECTOR MODE FIELD CONSTRUCTION AND DEGENERACIES VIA SYMMETRY

As noted in Sec. 2.7, the standard perturbation recipe for calculating eigenvalues to first-order and zeroth-order eigenstates for perturbed problems, such as the full VWE, effectively involves diagonalizing the RHS perturbation term. Without considerable intuition or hindsight, such calculations in general can be rather cumbersome. However, the situation can be simplified by using symmetry arguments as did Snyder and Young—see Sec. IIIB and footnote 13 of Ref. [48]. Group theory is the tool that formalizes their intuitive symmetry arguments; it is particularly effective when we go on to consider more complex situations. In this problem it provides us with a direct and elegant way of finding both the appropriate linear combination of pseudo-modes that form zeroth-order field $\tilde{\mathbf{e}}_t$ of Eq. (2.16*b*) for the true modes of the fiber and the degeneracies in the propagation constant splitting when the first-order eigenvalue correction $U^{(1)}$ is considered. As well as providing a formal and instructive derivation of standard construction with LP pseudo-modes, the method leads to the novel CP weak-guidance construction and classification [1] of the alternative set pioneered by Kapany and Burke [70] who considered full vector modes in terms of a circularly polarized basis [61, 70, 84].

3.3.1 Vector Field

In general we may write the transverse vector field $\mathbf{e}_t = e_x\hat{\mathbf{x}} + e_y\hat{\mathbf{y}} = e_+\hat{\mathbf{R}} + e_-\hat{\mathbf{L}}$ as

$$\mathbf{e}_t(\mathbf{r}) = \sum_{k=1}^{2} e_k(\mathbf{r})\hat{\mathbf{p}}_k = \mathbf{e}_t(\{\text{scalar}(\mathbf{r})\}, \{\text{polarization}\}) \tag{3.5}$$

For the modes of VWE$^{(0)}$, which have diagonal form with corresponding operator $\mathbb{H}_0$ being a multiple of the unit matrix, each field component magnitude $\tilde{e}_k(\mathbf{r})$ is simply a solution of the same SWE, $\mathcal{H}_0\psi_s = \tilde{U}^2\psi_s$; that is, a general field associated with propagation constant β_{lm} has the form

$$\tilde{\mathbf{e}}_t\ (\mathbf{r}) = \sum_{k=1}^{2} \tilde{e}_k(\mathbf{r})\hat{\mathbf{p}}_k \quad \text{where} \quad \tilde{e}_k(\mathbf{r}) = \sum_{s=1}^{2} a_s\psi_s(\mathbf{r}) = \sum_{s=1}^{2} a_s\psi_{lm}^{sk}(\mathbf{r}) \tag{3.6}$$

and we are free to choose any a_s. This corresponds to *arbitrary* linear combinations of the set of the degenerate modes

$$\tilde{\mathbf{e}}_t(\text{pseudo-mode}_{lm}^{sk}) = \tilde{\mathbf{e}}_t(\mathbf{r}) = \psi_s(\mathbf{r})\hat{\mathbf{p}}_k \tag{3.7}$$

However, if we also require that $\tilde{\mathbf{e}}_t$ be the zeroth-order term in the expansion of the true mode field solution of the exact VWE, then a_s is restricted so that $\tilde{\mathbf{e}}_t$ must be a *particular* linear combination and takes the form

$$\tilde{\mathbf{e}}_t(\text{true mode}_{\nu M}^{sk}) = \tilde{\mathbf{e}}_t(\mathbf{r}) = \sum_{k=1}^{2}\sum_{s=1}^{2} \mathbf{c}_{hsk}^{\nu}\psi_S(\mathbf{r})\hat{\mathbf{p}}_k \tag{3.8}$$

that is, *true VWE mode zeroth-order fields are specific combinations of pseudo-modes.* The pseudo-mode and true modal forms and in particular the coefficients $c_{hsk}^{\nu} = c_{hsk}^{\nu l}$ (which "couple" field magnitude to polarization) follow immediately upon recognition of the appropriate symmetries. Before giving the group theoretic origin of these coefficients in Table 3.4, we provide some intuitive background. In particular, we associate symmetry groups with VWE$^{(0)}$ and VWE, giving examples of their transformation properties. Analogous to the SWE case, the pseudo-and true modes are basis functions of the appropriate matrix representations.

In regard to the general form of a vector field in Eq. (3.5), it is convenient to first *associate separate symmetry groups with scalar magnitude and with polarization.*

3.3.2 Polarization Vector Symmetry Group: $\mathbf{C}^{\mathrm{P}}_{\infty\mathrm{v}}$

To represent the electric field polarization direction, for isotropic fibers, we are free to define a set of orthogonal axes $\{\hat{\mathbf{p}}_1, \hat{\mathbf{p}}_2\}$ that is in any direction and independent of position **r**; that is, the $\hat{\mathbf{p}}_k$ are simply parameterized by an arbitrary angle ϕ', and thus the polarization symmetry group (labeled by superscript P) is $\mathbf{C}^{\mathrm{P}}_{\infty\mathrm{v}}$ for all isotropic fibers independent of their transverse refractive index distribution. In particular, from Table A.4 we have $\{\hat{\mathbf{x}}, \hat{\mathbf{y}}\}$ forming a set of degenerate polarization directions that transforms as the real irrep $\mathbf{1}(\mathbf{C}^{\mathrm{P}}_{\infty\mathrm{v}})$, for example,

$$C_{\mathrm{P}}(\theta)\hat{\mathbf{x}} = \hat{\mathbf{x}}D^{(1)}(C_\theta)_{11} + \hat{\mathbf{y}}D^{(1)}(C_\theta)_{21} = \hat{\mathbf{x}}\cos\theta + \hat{\mathbf{y}}\sin\theta \tag{3.9}$$

Alternatively $\{\hat{\mathbf{R}}, \hat{\mathbf{L}}\}$ forms another set of degenerate polarization directions that transforms as the irrep **1** for a complex basis.

§ For anisotropic media the two polarization directions are no longer degenerate, and as we will see in Chap. 5, they transform as separate irreps of a lower symmetry group: (1) for *linear birefringence*, polarization symmetry is reduced to $\mathbf{C}^{\mathrm{P}}_{2\mathrm{v}}$ and the appropriate polarization directions $\hat{\mathbf{x}}$ and $\hat{\mathbf{y}}$ transform as the irreps $\mathbf{1}(\mathbf{C}^{P}_{2\mathrm{v}})$ and $\tilde{\mathbf{1}}(\mathbf{C}^{P}_{2\mathrm{v}})$; (2) for *radial birefringence*, $\hat{\mathbf{r}}$ and $\hat{\phi}$ transform as **0** and $\tilde{\mathbf{0}}$ of $\mathbf{C}^{\mathrm{P}}_{\infty\mathrm{v}}$ and for *circular birefringence*, $\hat{\mathbf{R}}$ and $\hat{\mathbf{L}}$ transform as **+1** and **–1** of $\mathbf{C}^{\mathrm{P}}_{\infty}$. §

> *General polarization group definition:* In general, corresponding to a group **G** with group elements g, we *define* the polarization symmetry group $\mathbf{G}^{\mathrm{P}}$ to be the group of operators $g_{\mathrm{P}} \equiv O_{\mathrm{P}}(g)$, which act specifically on the polarization basis vectors $\hat{\mathbf{p}}_k$.

3.3.3 Zeroth-Order Vector Wave Equation Symmetry: $\mathbf{C}^{\mathrm{S}}_{\infty\mathrm{v}} \otimes \mathbf{C}^{\mathrm{P}}_{\infty\mathrm{v}}$

We saw in Sec. 2.7 that the operator $\mathbb{H}_0$ corresponding to VWE$^{(0)}$ is diagonal and a multiple of the unit matrix, and thus places no restriction on the relative magnitudes of the two components $\tilde{e}_k$ (**r**), which independently satisfy the same SWE with operator H_{S}. Hence the field polarization direction is uncoupled to magnitude and thus position. The appropriate $VWE^{(0)}$ symmetry group is simply the direct product (denoted $\otimes$) of the scalar and polarization symmetry groups, that is, $\mathbf{G}^S \otimes \mathbf{G}^P$. In particular,

1. For **isotropic** fibers, in general we have

$$\boxed{\mathbb{H}_0 - \text{symmetry} = (\mathcal{H}_S - \text{symmetry}) \otimes \mathbf{C}^P_{\infty v}\,(\text{general, isotropic})} \quad (3.10a)$$

2. For **circular** isotropic fibers this reduces to

$$\boxed{\mathbb{H}_0 - \text{symmetry} = \mathbf{C}^S_{\infty v} \otimes \mathbf{C}^P_{\infty v} \quad (\text{circular, isotropic})} \quad (3.10b)$$

Symmetry Operations: Definition These correspond to independent action of the operators of the two groups. Because it is a direct product, we can choose all possible combinations $O(g, h) \equiv O_S(g)O_P(h) \equiv g_S h_P$ such that

$$O(g,h)\tilde{\mathbf{e}}_t(\mathbf{r}) = \sum_{k=1}^{2} \{O_S(g)\tilde{\mathbf{e}}_k(\mathbf{r})\}\{O_p(h)\hat{\mathbf{p}}_k\} \qquad g,h \in \mathbf{C}_{\infty v} \quad (3.11a)$$

or equivalently

$$g_S h_P \tilde{\mathbf{e}}_t(\mathbf{r}) = \sum_{k=1}^{2} \{g_S \tilde{\mathbf{e}}_k(\mathbf{r})\}\{h_P \hat{\mathbf{p}}_k\} \qquad g_S \in \mathbf{C}^S_{\infty v}, h_P \in \mathbf{C}^P_{\infty v} \quad (3.11b)$$

with obvious generalization to the case of scalar and polarization groups $\mathbf{G}^S$ and $\mathbf{H}^P$ being associated with different basic groups $\mathbf{G}$ and $\mathbf{H}$.

Symmetry Operations: Examples To illustrate the effect of the above symmetry operations on the vector field, consider the following examples involving the vector modes introduced in Sec. 3.1. As the scalar and polarization actions are independent, it suffices to consider them separately, i.e., the operations $O_S(g) = O(g, E)$ and $O_P(h) = O(E, h)$.

As for the scalar modes, **conversion from an even to an odd LP mode** is obtained via a **scalar rotation of $\pi/2l$**, for example, by $\pi/4$ for conversion of LP_{21}^{ex} to LP_{21}^{ox}.

$$C_S\left(\tfrac{\pi}{4}\right)LP_{21}^{ex} = C_S\left(\tfrac{\pi}{4}\right)\!\left(\!\begin{array}{c}\ominus\ \oplus\\\oplus\ \ominus\end{array}\!\right) = \{C_S\left(\tfrac{\pi}{4}\right)\!\left(\!\begin{array}{c}\ominus\ \oplus\\\oplus\ \ominus\end{array}\!\right)\}\,\{\mathbf{t}\} = \{\left(\!\begin{array}{c}\ominus\ \oplus\\\oplus\ \ominus\end{array}\!\right)\}\,\{\mathbf{t}\} = \left(\!\begin{array}{c}\oplus\ \oplus\\\ominus\ \ominus\end{array}\!\right) = LP_{21}^{ox}$$

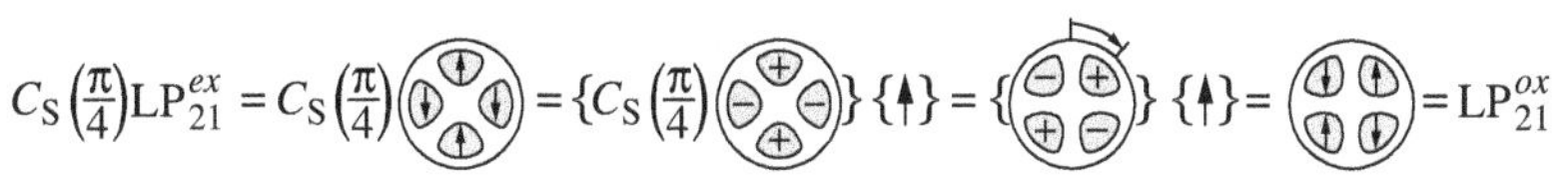

$$(3.12a)$$

By contrast, **polarization rotation by** $\pi/2$ is required for **conversion from** x– **to** y–**polarized LP modes** independent of their azimuthal order, e.g., for LP_{21} we have

$$C_P\left(\tfrac{\pi}{2}\right)LP_{21}^{ex} = C_P\left(\tfrac{\pi}{2}\right)\!\!\bigcirc\!\! = \{\bigcirc\}\,\{C_P\left(\tfrac{\pi}{2}\right)\!\!\uparrow\!\!\} = \{\bigcirc\}\,\{\rightarrow\} = \bigcirc = LP_{21}^{ey}$$

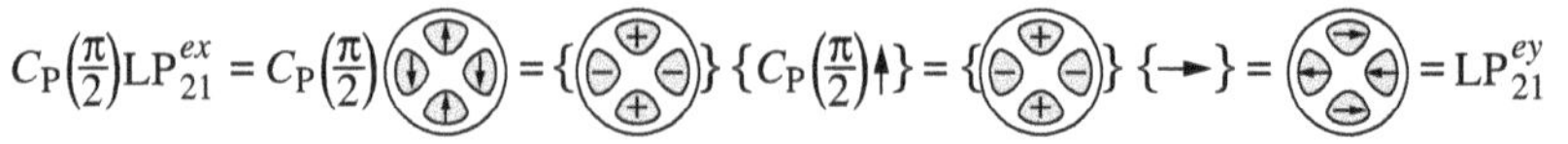

$$(3.12b)$$

In general, we find that scalar rotation acts on the LP modes as

$$C_S(\theta)LP_{lm}^{ex} = LP_{lm}^{ex}\cos l\theta + LP_{lm}^{oy}\sin l\theta \qquad (3.12c)$$

whereas polarization rotation gives

$$C_P(\theta)LP_{lm}^{ex} = LP_{lm}^{ex}\cos\theta + LP_{lm}^{oy}\sin\theta \qquad (3.12d)$$

Thus, as expected from our previous discussion of scalar mode transformations and polarization vector transformations, the LP_{lm} modes transform as the irrep **1** under scalar rotations and as the irrep **1** under polarization rotations. The same result holds for arbitrary linear combinations of these modes.

3.3.4 Pseudo-Vector Modes: Basis Functions of Irreps of $C_{\infty V}^S \otimes C_{\infty V}^P$

The consequence of the above discussion is that the appropriate irreps that give the transformation properties of the modes of $VWE^{(0)}$ are simply the products of the irreps **1** corresponding to the SWE modes and the irrep **1** corresponding to polarization; i.e., we need to consider the irrep products $\mathbf{1} \otimes \mathbf{1}$ of the direct group product $C_{\infty v} \otimes C_{\infty v}$ or more explicitly $l \otimes 1(C_{\infty v}^S \otimes C_{\infty v}^P) = l(C_{\infty v}^S) \otimes 1(C_{\infty v}^P) \equiv l_{\infty v}^S \otimes 1_{\infty v}^P$.

In particular, taking the set $\Psi_s^{(l)} \in F_l(R)\{\cos l\phi, \sin l\phi\}$ as a basis (b) for the irreps l of the SWE magnitude group $C_{\infty v}^S$, and $\hat{p}_k \in \{\hat{x}, \hat{y}\}$ as a basis for the irrep **1** of the polarization group $C_{\infty v}^P$, we obtain the irrep product basis functions corresponding to the LP_{lm} **modes** of Table 3.1, i.e.,

$$\boxed{LP_{lm}^{sk} = \left[\Psi_s^{(l)}\hat{p}_k\right]_{b=\text{real}} \in F_l(R)\{\cos l\phi\hat{x}, \cos l\phi\hat{y}, \sin l\phi\hat{x}, \sin l\phi\hat{y}\}} \qquad (3.13a)$$

Similarly, choosing a complex exponential basis (b = complex) and taking the set $\Psi_{\mathbf{s}}^{(l)} \in F_l(R)\{e^{il\phi}, e^{-il\phi}\}$ as a basis for the irreps $\mathbf{l}(\mathbf{C}_{\infty v}^{S})$, and $\hat{\mathbf{p}}_k \in \{\hat{\mathbf{R}}, \hat{\mathbf{L}}\}$ as a basis for the irrep $\mathbf{1}(\mathbf{C}_{\infty v}^{P})$, we obtain the irrep product basis functions corresponding to the $\mathbf{CP}_{lm}$ **modes** of Table 3.1, i.e.,

$$\boxed{\mathrm{CP}_{lm}^{sk} = \left[\Psi_{\mathbf{s}}^{(l)}\hat{\mathbf{p}}_k\right]_{b=\text{complex}} \in F_l(R)\{e^{il\phi}\hat{\mathbf{R}},\ e^{-il\phi}\hat{\mathbf{R}},\ e^{il\phi}\hat{\mathbf{L}},\ e^{il\phi}\hat{\mathbf{L}}\}} \quad (3.13b)$$

In both cases, the **number of degenerate propagation constants**, given by the dimension of the irrep product $\boldsymbol{l} \otimes \mathbf{1}$, is 2 for azimuthal order $l = 0$ and 4 for $l \geq 1$.

3.3.5 Full Vector Wave Equation Symmetry: $\mathbf{C}_{\infty v}^{S} \otimes \mathbf{C}_{\infty v}^{P} \supset \mathbf{C}_{\infty v}^{J}$

As discussed in Sec. 2.7, inclusion of finite-Δ effects in the vector wave equation via $\mathbb{H}_{\mathbf{p1}}$ of Eq. (2.19) (which in contrast to $\mathbb{H}_0$ is nondiagonal) results in a coupling of the field polarization components or equivalently a coupling of polarization direction to scalar magnitude and thus to position, as can be understood from Eq. (3.8). In terms of symmetry, inclusion of the $\mathbb{H}_{\mathbf{p1}}$ means that the vector wave equation is no longer invariant under rotation-reflection acting on magnitude or polarization independently, i.e., for general $O(g, h)$ defined by Eq. (3.11), $O(g, h)\,\mathbb{H}_{\mathbf{p1}}\,\mathbf{e}_t \neq \mathbb{H}_{\mathbf{p1}} O(g, h)\,\mathbf{e}_t$.

However, it can be shown [1] that $\mathbb{H}_{\mathbf{p1}}$ and thus VWE$^{(1)}$ (and in fact $\mathbb{H}_{\mathrm{pol}}$ and VWE) are invariant under **joint** rotation (or reflection) of magnitude and polarization. This corresponds to invariance under the subset of symmetry operations $O(g, h)$ defined by Eq. (3.11) that has $g = h$, that is,

$$g_J\,\mathbb{H}_{\mathbf{p1}}\,\mathbf{e}_t = \mathbb{H}_{\mathbf{p1}}\,g_J\,\mathbf{e}_t,$$

$$\text{where} \qquad g_J \equiv O(g, g) = O_S(g)O_P(g) \in \mathbf{C}_{\infty v}^{J},\ g \in \mathbf{C}_{\infty v} \qquad (3.14)$$

Thus, $\mathbf{C}_{\infty v}^{J}$ consisting of *diagonal* operators $O(g, g)$ is defined as the *diagonal subgroup* of $\mathbf{C}_{\infty v}^{S} \otimes \mathbf{C}_{\infty v}^{P}$. We denote restriction of the operations of a group $\mathbf{G}$ to those of a subgroup $\mathbf{G}_s$ by $\mathbf{G} \supset \mathbf{G}_s$ and thus denote the symmetry reduction due to inclusion of $\mathbb{H}_{\mathbf{p1}}$ in the VWE by $\mathbf{C}_{\infty v}^{S} \otimes \mathbf{C}_{\infty v}^{P} \supset \mathbf{C}_{\infty v}^{J}$.

Understanding the Symmetry Operations With regard to mode forms, joint symmetry operations correspond to rotation (or reflection) of the mode pictogram as a whole.

For example, considering the HE_{21} mode defined in Table 3.2, we have

$$C_J\left(\frac{\pi}{4}\right) HE^{e}_{2m} = C_J\left(\frac{\pi}{4}\right)$$ 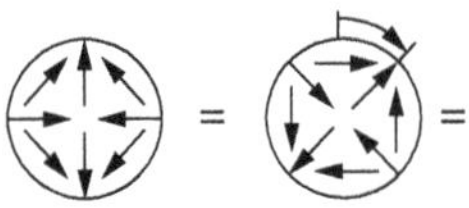 $$= HE^{o}_{2m} \qquad (3.15)$$

(Note that to show these modes are identical within a rotation, we have given the field direction at azimuthal positions ϕ as integer multiples of $\pi/4$ rather than just $\pi/2$ multiples for the simplified pictograms.)

For TE_{01}, also defined in Table 3.2, rotation of the pictogram has no effect; i.e., joint rotations leave TE_{01} invariant:

$$C_J\left(\frac{\pi}{2}\right) TE_{01} = C_J\left(\frac{\pi}{2}\right)$$ 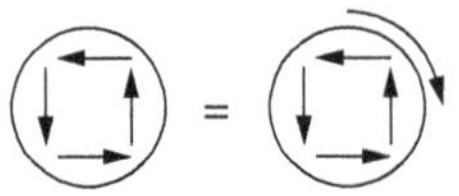 $$= TE_{01} \qquad (3.16)$$

However, if there is polarization rotation alone [i.e., rotation of the polarization direction of the electric field at each point $\mathbf{r} = (x, y)$ locally about that point independently of position], then TE_{01} converts to TM_{01}, or

$$C_P\left(\frac{\pi}{2}\right) TE_{01} = C_P\left(\frac{\pi}{2}\right)$$ 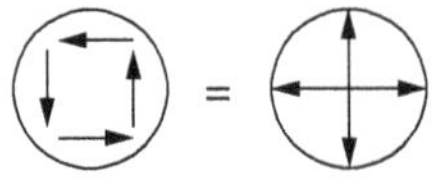 $$TM_{01} \qquad (3.17)$$

As we will see more explicitly in the next section, these transformations have significance regarding mode splitting and degeneracies. In particular, the conversion between the two HE_{21} modes indicates degeneracy even when the symmetry is restricted to joint operations, as is the case when $\mathbb{H}_{\mathbf{p1}}$ is considered. However, the conversion of TE_{01} under joint rotations to itself indicates nondegeneracy when $\mathbb{H}_{\mathbf{p1}}$ is considered. By contrast if separate operations were allowed, then the conversion between TE_{01} and TM_{01} would imply degeneracy of these modes: For a general radial profile, this is only the case for modes of the $VWE^{(0)}$ that are $\mathbf{C}^{\mathbf{S}}_{\infty\mathbf{v}} \otimes \mathbf{C}^{\mathbf{P}}_{\infty\mathbf{v}}$ invariant (and for which TM/TE/HE/EH modes are degenerate solutions as equally valid as LP modes).

3.3.6 True Vector Modes: Qualitative Features via $\mathbf{C}^{\mathbf{S}}_{\infty\mathbf{v}} \otimes \mathbf{C}^{\mathbf{P}}_{\infty\mathbf{v}} \supset \mathbf{C}^{\mathbf{J}}_{\infty\mathbf{v}}$

Before giving, in Table 3.4, the formal construction of the true VWE modes as required by the symmetry reduction $\mathbf{C}^{\mathbf{S}}_{\infty\mathbf{v}} \otimes \mathbf{C}^{\mathbf{P}}_{\infty\mathbf{v}} \supset \mathbf{C}^{\mathbf{J}}_{\infty\mathbf{v}}$, we first discuss qualitative features: branching rules, resulting level splitting, and modal transformation properties.

Branching Rules

When symmetry operations are restricted to a subgroup, the matrix representations of the restricted symmetry operations can be decomposed as a direct sum of the irrep matrices of the subgroup. These are given by standard branching rules as described in Sec. A.3. The symmetry reduction $\mathbf{C}_{\infty v} \otimes \mathbf{C}_{\infty v} \supset \tilde{\mathbf{C}}_{\infty v}$ is a direct-product reduction, and knowledge of how the representations of $\mathbf{C}_{\infty v} \otimes \mathbf{C}_{\infty v}$ decompose into irreps of $\mathbf{C}_{\infty v}$ (as in the second column in the lower part of Table 3.3) provides (1) a determination of the qualitative splitting and degeneracies of the propagation constant β when going from pseudo-to true vector modes and (2) the transformation properties of the resulting modes, allowing an association with irreps if we already know their form. These two features are explained, respectively, in the next two subsections.

Polarization Level Splitting—Qualitative Features via Irrep Dimensions

Because the product $\boldsymbol{l}^{\mathrm{S}}_{\infty v} \otimes \mathbf{1}^{\mathrm{P}}_{\infty v}$ is reducible into irreps $\nu^{\mathrm{J}}_{\tilde{\infty} v}$ the level associated with the propagation constant $\tilde{\beta}_{lm}$ (which has degeneracy equal to the corresponding irrep dimension $|\boldsymbol{l} \otimes \mathbf{1}|$) is split into n *sublevels* β^{ν}_{lm} with degeneracy $|\nu|$. For example, from the second column of Table 3.3, it is seen that for the $\boldsymbol{l} = \mathbf{1}$ reduction, the fourfold degenerate level corresponding to $\mathbf{1}^{\mathrm{S}}_{\infty v} \otimes \mathbf{1}^{\mathrm{P}}_{\infty v}$ generates three sublevels transforming as irreps $\nu^{\mathrm{J}}_{\infty v}$ with $\boldsymbol{\nu} = \mathbf{0}, \tilde{\mathbf{0}}$, and $\mathbf{2}$, which are of dimensions 1, 1, and 2, respectively. Thus an immediate consequence of symmetry is that when one accounts for finite Δ, for modes of azimuthal order $l = 1$, one expects two modes with nondegenerate propagation constants corresponding to the irreps $\mathbf{0}$ and $\tilde{\mathbf{0}}$, and two modes with degenerate propagation constants corresponding to the irrep $\mathbf{2}$.

Association of True Modes with Representations

Although formal construction of true modes associated with irreps $\boldsymbol{\nu}$ is given in the next subsection, if we already know the form of the modal fields, as in Fig. 3.1, we can identify each mode with a particular representation by examination of the transformation resulting from each symmetry operation.

For example, for TM_{0m}, all symmetry operations of $\mathbf{C}_{\infty v}$ leaving the field $\tilde{\mathbf{e}}_t$-invariant indicate that the identity representation **0** in Table A.4 is appropriate. However, although rotations leave the TE_{0m} electric field–invariant, reflection inverts it: this corresponds to $\mathbb{D}^{(0)}(C_\theta) = 1$ and $\mathbb{D}^{(\tilde{0})}(\sigma_v) = -1$ in Table A.4: thus we identify the representation $\tilde{\mathbf{0}}$ with this mode.

For $HE^{e}_{\nu m}$ it may be shown that a rotation by angle θ acting jointly on magnitude and the polarization leads to a mixing $HE^{e}_{\nu m}\cos\nu\theta - HE^{o}_{\nu m}\sin\nu\theta$. The deduction from this is that these two HE modes are degenerate and transform according to the irrep ν. Similar results hold for all hybrid modes. A detailed demonstration of the irrep identification of each mode is given in Ref. [94].

3.3.7 True Vector Modes via Pseudo-Modes: Basis Functions of $\mathbf{C}^{S}_{\infty v} \otimes \mathbf{C}^{P}_{\infty v} \supset \mathbf{C}^{J}_{\infty v}$

It is the symmetry reduction $\mathbf{C}^{S}_{\infty v} \otimes \mathbf{C}^{P}_{\infty v} \supset \mathbf{C}^{J}_{\infty v}$, that is, the restriction to joint symmetry operations, that restricts the zeroth-order term $\mathbf{e}_t$ in the expansion of the true VWE mode fields $\mathbf{e}_t$ (assuming finite but small Δ) to be the specific combinations of the $VWE^{(0)}$ pseudo-modes mentioned in Sec. 3.3.1. Group theoretically, these combinations correspond to the basis functions of $\mathbf{C}^{J}_{\infty v}$ which are directly obtained via a standard transformation [7, 9] as linear combinations of the product basis functions of $\mathbf{C}^{S}_{\infty v} \otimes \mathbf{C}^{P}_{\infty v}$. This transformation in fact is just Eq. (3.8) with the basis transformation coefficients $c^{\nu l}_{hsk}$ that give the linear combinations [or equivalently that couple the scalar basis functions $\Phi^{l}_{s}(\phi)$ to the polarization directions $\hat{\mathbf{p}}_k$] being the standard Clebsch-Gordan coupling coefficients $c^{\nu l 1}_{hsk} \equiv \langle ls, 1k \mid \nu h \rangle$ described in Sec. A.4.

The resulting $\mathbf{C}^{J}_{\infty v}$ basis functions $\hat{\mathbf{p}}^{\nu l}_{h}(\phi)$, which give the azimuthal dependence of $\tilde{\mathbf{e}}_t$ for the true modes, are summarized in part (*c*) of Table 3.4; i.e., the transformation takes the form

$$[\tilde{\mathbf{e}}_t(\text{true mode}^{sk}_{\nu m})]_b = F_{lm}(R)[\hat{\mathbf{p}}^{\nu l}_{h}(\phi)]_b$$
$$= \sum_{k=1}^{2}\sum_{s=1}^{2} [c^{\nu l 1}_{hsk} \quad \tilde{\mathrm{e}}_t(\text{pseudo-mode}^{sk}_{lm})]_b \qquad (3.18)$$

where b indicates that the expression may be evaluated in a real $(b = r)$ or complex $(b = c)$ basis, and the mode labels are summarized at the bottom of Table 3.4.

Note that the true mode polarization dependence on azimuthal angle $\hat{\mathbf{p}}_h^{\nu l}(\phi)$ is completely determined by the symmetry properties of $\mathbb{H}_{\mathbf{p}1}$, that is, the fact that it is invariant under joint rotation-reflection, rather than its details.

§ Note also that the basis functions $\hat{\mathbf{p}}_h^{\nu l}(\phi)$ of $\mathbf{C}_{\infty\mathbf{v}}^{\mathrm{J}}$ are just the vector basis functions (VBFs) of $\mathbf{C}_{\infty\mathbf{v}}$ given in the right-hand column of Table A.4 and applicable to an arbitrary physical system with the given symmetry. They may be described as vector cylindrical harmonics in analogy with the well-known vector spherical harmonics given in Ref. 95. §

Standard True Vector Mode Construction with LP Modes

Substituting the LP modes with the real basis CG coefficients, i.e., in Eq. (3.18), gives the standard true vector mode set (left section of Table 3.2).

Alternative True Vector Mode Construction with CP Modes

Similarly, substituting the CP modes with the complex basis CG coefficients in Eq. (3.18) gives the standard true vector mode set (right section of Table 3.2).

We reiterate that, unlike the LP modes, which are all pseudo-modes for $l > 0$, *the circularly polarized CP modes are true weak-guidance vector modes* that are equally valid as the hybrid even/odd polarized $\mathrm{HE}_{\nu m}^{e/o}$ and $\mathrm{EH}_{\nu m}^{e/o}$ modes, *except for* CP_{1m}^{R+} and CP_{1m}^{L-} which are only pseudo-modes. These exceptions correspond to azimuthal mode number $\nu = 0$ for which the only true modes are TM_{0m} and TE_{0m}. As these latter two modes are nondegenerate (corresponding to different irreps $\mathbf{0}$ and $\tilde{\mathbf{0}}$), unlike modes with $\nu > 0$, they cannot be combined to obtain circularly polarized true mode forms. A schematic interpretation of CP mode field evolution along a fiber is given in Fig. 3.2.

3.4 POLARIZATION-DEPENDENT LEVEL-SPLITTING

3.4.1 First-Order Eigenvalue Corrections

For circular symmetry the first-order eigenvalue corrections given by Eq. (2.23) can be evaluated by noting the separation

$$\mathsf{H}_{jj} = <\tilde{\mathbf{e}}_t, \mathbb{H}_{\mathbf{p}1}\tilde{\mathbf{e}}_t> = I_{\phi 1} I_{r1} + I_{\phi 2} I_{r2} \tag{3.19}$$

where the radial integrals are

$$I_{r1,lm} = \int_0^\infty F_{lm} \frac{dF_{lm}}{dR} \frac{df}{dR} dR \qquad \text{and} \qquad I_{r2,lm} = \int_0^\infty F_{lm}^2 \frac{df}{dR} dR \tag{3.20}$$

and the angular integrals are

$$I_{\phi 1,\,lmp} = \pi\{1 + (\delta_{p2} - \delta_{p4})\delta_{l1}\} = \pi\{1 + \delta_{v0} - \delta_{v\tilde{0}}\} \tag{3.21a}$$

and

$$I_{\phi 2,\,lmp} = l\pi\{(-1)^p + (\delta_{p2} - \delta_{p4})\delta_{l1}\} = l\pi\{v - l + \delta_{v0} - \delta_{v\tilde{0}}\} \tag{3.21b}$$

We remark that grouped theoretically this quantification of the splittings corresponds to extracting the symmetry information from H_{jj} via the Wigner-Eckart theorem [9]. In particular this says that the *radial integrals,* which correspond to reduced matrix elements, are *independent of polarization;* the *angular integrals* contain the $\mathbf{C}_{\infty \mathbf{v}} \otimes \mathbf{C}_{\infty \mathbf{v}} \supset \mathbf{C}_{\infty \mathbf{v}}$ symmetry reduction information and *are proportional to Clebsch-Gordan coefficients.*

Evaluation of the angular integrals (Clebsch-Gordan coefficients) leads to the explicit splittings to first order in Δ in terms of the radial integrals (reduced matrix elements) in Table 3.5. There we give the general forms of the circular fiber polarization splittings for both general and example radial dependencies of the profile, as discussed in the following subsections. To reduce the mode labels, as in Ref. 3 and the fourth column of Table 3.2, we use the single polarization mode number p to correspond to the natural irrep symmetry labeling vh.

3.4.2 Radial Profile-Dependent Polarization Splitting

By variation of the radial dependence of the profile, one may manipulate the magnitude and sign of the modal polarization splittings. For example, we will see in Fig. 3.3 that for step profiles, the TM_{01} mode has a larger eigenvalue U (smaller effective index) than has TE_{01}; however, for triangular ($q = 1$) grading of the profile,

TABLE 3.5

Normalized First-Order Eigenvalue Corrections for Circular Core Fibers

				$H_{jj} \equiv \{2\tilde{U}N/\pi\}U^{(1)}_{lmp}$		
$C^S_{\infty v}$ irrep l	$C^J_{\infty v}$ irrep ν	h	**Circular Fiber Mode**	**General Circular Profile** $f = f(R)$	**Infinite Parabolic** $f = R^2, \forall R$	**Infinite Power Law** $f = R^q, \forall R$
$l = 0$: $p = 1, 3$	**1**	1, 2	HE^e_{1m}, HE^o_{1m}	$I_{r1,0m}$	$I_{r1,0m}$	$I_{r1,0m}$
$l = 1$:						
$p = 2$	**0**	1	TM_{0m}	$2(I_{r1,lm} + I_{r2,1m})$	0	$2\left(1-\frac{2}{q}\right)I_{r1,lm}$
$p = 4$	$\tilde{\mathbf{0}}$	1	TE_{0m}	0	0	0
$p = 1, 3$	**2**	1, 2	HE^e_{2m}, HE^o_{2m}	$I_{r1,1m} - I_{r2,1m}$	$2I_{r1,1m}$	$\left(1+\frac{2}{q}\right)I_{r1,lm}$
$l > 1$:						
$p = 2, 4$	$l-1$	1, 2	EH^e_{l-1m}, EH^o_{l-1m}	$I_{r1,lm} + l\, I_{r2,lm}$	$(1-l)I_{r1,lm}$	$\left(1-\frac{2l}{q}\right)I_{r1,lm}$
$p = 1, 3$	$l+1$	1, 2	HE^e_{l+1m}, HE^o_{l+1m}	$I_{r1,lm} - l\, I_{r2,lm}$	$(1+l)I_{r1,lm}$	$\left(1+\frac{2l}{q}\right)I_{r1,lm}$

As defined in Eq. (3.19) in terms of the integrals I of Eqs. (3.20) and (3.21), H_{jj} gives the first-order eigenvalue corrections $U^{(1)}_{lmp}$ defined by Eq. (2.23) within a factor $\{2\tilde{U}N/\pi\}$, where N is the field normalization of Eq. (2.24).

FIGURE 3.3

TM_{01}-TE_{01} polarization eigenvalue splitting for step and clad power-law profiles [defined by $f(R) = R^q$, for $R \leq 1$ and $f(R) = 1$ for $R > 1$]. Normalization is with respect to the profile height parameter Δ and the zeroth-order eigenvalue $\tilde{U}$. Note that this splitting is given by the TM_{01} eigenvalue correction, that is, $\delta U = \Delta U_{112}^{(1)}$, because the TE_{01} eigenvalue correction $U_{114}^{(1)}$ is always 0. For the *clad-parabolic profile* with large V the TM_{01} and TE_{01} modes become degenerate; this corresponds to the *infinite parabolic profile* for which these modes are degenerate for all V. For the *step profile* these modes are degenerate at cutoff. However, for the power-law profiles ($q < \infty$), (1) cutoff for TM_{01} occurs at a lower V value than for TE_{01}, and (2) if $2 < q < \infty$, TM_{01}-TE_{01} degeneracy occurs at one V value, which is increasingly farther from cuttoff as q diminishes. Although the first-order perturbation expression for the eigenvalue correction gives no splitting at cutoff even for $q < \infty$, that is, $U_{112}^{(1)}$ does approach 0, note that first-order perturbation theory becomes inaccurate very close to cutoff (dashed lines), as there $\tilde{\mathbf{e}}_t$ (which corresponds to the TE_{01} field) becomes a bad approximation for the exact TM_{01} field $\mathbf{e}_t$. Rather than the eigenvalue correction in this region, the weak-guidance perturbation approach can be adapted to obtain the V value for TM_{01} cutoff. Consideration of higher-order terms may useful for increasing the accuracy [96]; cf. Sec. 8.5 of Ref. 97 where the cutoff formulation is used for finite-cladding fibers to obtain the profile dependence of HE_{11} core-mode cutoff, i.e., the V value at which $n_{\text{eff}} = n_{\text{cl}}$.

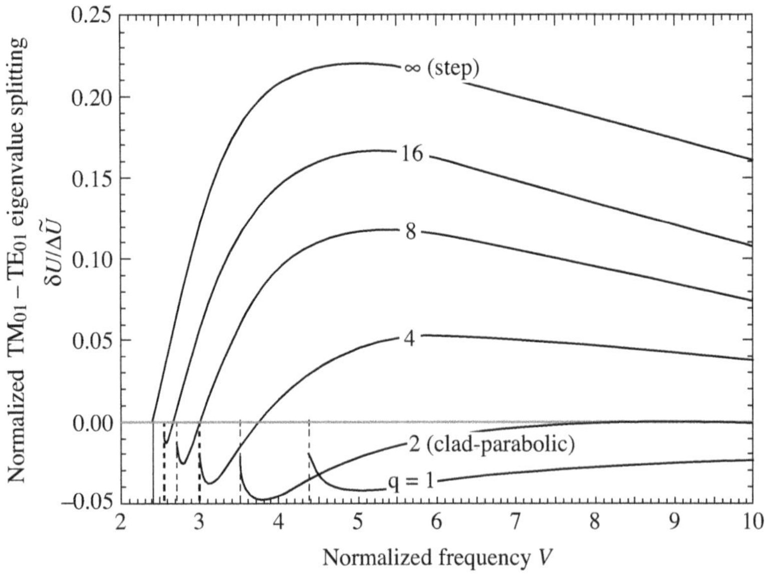

the reverse is the case [52, Chap. 4]. Such radial dependent profile splitting effects provide a macroscopic building block for the radial anisotropy discussed in Chap. 5.

3.4.3 Special Degeneracies and Shifts for Particular Radial Dependence of Profile

Given degeneracies due to circular symmetry, further degeneracy is possible for *special choices of radial dependence of profile.*

Infinite-Power-Law Profiles: $f(R) = R^q$, $0 \le R \le \infty$

These profiles have the symmetry property $(Rf')' = qR^q$; thus the radial integrals are related by $I_{r2,1m} = -(2l/q)I_{r1,1m}$, giving the shifts in Table 3.5. Note the special cases:

1. $q = 2$: Infinite parabolic, i.e., *harmonic oscillator* profile: For $l = 1$, the radial integrals cancel; that is, $I_{r2,1m} = -I_{r1,1m}$, giving $U^{(1)}_{1m2} = 0 (= U^{(1)}_{1m4})$, that is, *$TM_{0m}$ and TE_{0m} are first-order degenerate.*
2. $q = 6$: *TM_{0m} and HE_{2m}* modes *are first-order degenerate* with normalized shifts given by $(4/3)I_{r1,1m}$.
3. *$EH_{l-1,m}$ modes with $l = q/2$ undergo zero first-order shift;* i.e., their first-order eigenvalues are degenerate with those of the zeroth order or LP_{lm} modes.

Clad-Power-Law Profiles: $f(R) = R^q$, $0 \le R \le 1$; $f(R) = 1$, $1 \le R \le \infty$

For V much above a mode's cutoff, its field is essentially confined to the core and thus behaves as for an infinite-power-law profile. For large V, the infinite profile provides an excellent approximation for eigenvalue corrections. In particular, we see (1) in Fig. 3.3 that the above-mentioned infinite parabolic degeneracy results in the TM_{01}-TE_{01} splitting becoming negligible as V increases and (2) in Fig. 3.4 a similar effect occurs for TM_{01}-HE_{21} splitting in the case of $q = 6$ profiles.

In Fig. 3.3, we also see that another interesting TM_{01}-TE_{01} degeneracy occurs for step profiles ($q = \infty$) at cutoff. For power-law profiles with $\infty > q > 2$, this becomes the accidental degeneracy that occurs at V values increasingly above cutoff as q diminishes toward 2. In Fig. 3.4, we see that such an accidental degeneracy

FIGURE 3.4

TM_{01}-HE_{21} polarization eigenvalue splitting for step and clad power–law profiles (as in Fig. 3.3). Normalization is with respect to the profile height parameter Δ and the zeroth-order eigenvalue $\tilde{U}$. For the $q = 6$ *clad power-law profile* with large V, the TM_{01} and HE_{21} modes become degenerate; this corresponds to the $q = 6$ *infinite power-law profile* [i.e., $f(R) = R^6$, $0 \le R < \infty$] for which these modes are degenerate for all V. For the *step profile* these modes are degenerate at $V \approx 3.8$. For the clad power-law profiles with $\infty > q > 6$, this degeneracy occurs at increasing values of V as the grading parameter q diminishes toward 6.

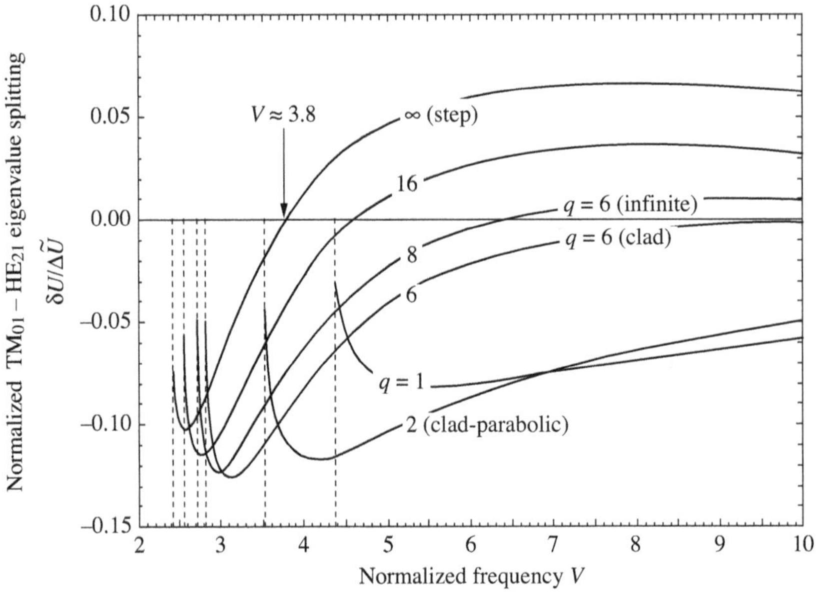

occurs between TM_{01} and HE_{21} eigenvalues for $V \approx 3.8$ for step profiles and increasing V values as the power-law grading parameter diminishes toward $q = 6$. If finite or multiple claddings are considered, these degeneracies may occur at multiple V values.

3.4.4 Physical Effects

Eigenvalue Splitting and Polarization Beating

In general, the excitation of a linear polarized pseudo-mode corresponds to the excitation of two true modes; e.g., from Table 3.2, we obtain

$$\{LP_{11}^{ex}, LP_{11}^{oy}\} = TM_{01} \pm HE_{21}^{e} \qquad \text{radial LP polarization} \qquad (3.22a)$$

$$\{LP_{11}^{ox}, LP_{11}^{ey}\} = HE_{21}^{o} \pm TE_{01} \qquad \text{azimuthal LP polarization} \qquad (3.22b)$$

Eigenvalue splitting between the two constituent true modes (e.g. between TM_{01} and HE_{21}^{e} or between TE_{01} and HE_{21}^{o}) thus results in polarization rotation. In general the splittings and thus beat lengths for the two cases differ. However, for cases of degeneracy between TM_{01} and TE_{01}, the resulting beat lengths are independent of whether the LP polarization is radial or azimuthal.

Complementary Polarization Degeneracies and LP True Modes

In cases of degeneracy between two true modes, any linear combination is also a true mode. Degeneracies between two *constituent* true modes mean that the *two associated* LP modes are also true modes that thus propagate undisturbed on an ideal fiber.

1. *Step profiles.* The **accidental degeneracies** at the particular V values for TM_{0m} and HE_{2m} mean that the corresponding radially polarized pair of LP modes LP_{1m}^{ex} and LP_{1m}^{oy} are also true modes at these V values, e.g., at $V \approx 3.8$ for TM_{01} and HE_{21} for which LP_{11}^{ex} and LP_{11}^{oy} are thus also true modes. Such V values have particular significance as transition points for the limiting forms of guides that are adiabatically deformed, as we will see in the following chapters.
2. $q = 6$. For clad $q = 6$ profiles, although exact TM_{0m}-HE_{2m} degeneracy only occurs in the limit $V \to \infty$, the TM_{0m}-HE_{2m} beat lengths become increasingly large as V increases, and thus LP_{1m}^{ex} or LP_{1m}^{oy} propagates for increasingly longer distances before complete polarization rotation occurs. That is, the LP modes can be regarded as true modes if length scales considered are much smaller than the corresponding TM_{0m}-HE_{2m} beat length.

CHAPTER 4

Azimuthal Symmetry Breaking

In this chapter, we consider examples of reduction of the geometrical symmetry of the fiber from circularity, as described by the continuous rotation-reflection group $\mathbf{C}_{\infty v}$, to discrete n-fold rotation-reflection symmetry corresponding to the group $\mathbf{C}_{nv}$. We examine the resulting propagation constant splitting as well as the transformation of modal fields resulting therefrom.

As particular examples, we discuss (1) $\mathbf{C}_{2v}$, the reduction of a circular to an elliptical fiber in detail illustrating the method (this is of particular interest in that it allows us to clarify some previous results concerning the transformation of vector modes corresponding to circular fiber modes of azimuthal symmetry $l = 1$), (2) $\mathbf{C}_{3v}$, the equilateral triangular deformation, and (3) $\mathbf{C}_{4v}$, the square deformation.

4.1 PRINCIPLES

4.1.1 Branching Rules

We have seen that both the scalar and the true vector modes of a circular guide are associated with irreps of $\mathbf{C}_{\infty v}$ (or more explicitly $\mathbf{C}^{S}_{\infty v}$ and $\mathbf{C}^{J}_{\infty v}$, respectively). To take account of a geometric perturbation of symmetry $\mathbf{C}_{nv}$ in both these cases, we need to consider branching rules for the symmetry reduction $\mathbf{C}_{\infty v} \supset \mathbf{C}_{nv}$ as in Sec. A.3. The VWE$^{(0)}$ modes of the geometrically perturbed fiber are simply given as the direct product of the scalar modes with polarization, which for an isotropic fiber is as for the circular case; i.e., polarization basis vectors, which are independent of fiber geometry and correspond to $\mathbf{1}(\mathbf{C}^{\mathbf{P}}_{\infty\mathbf{v}})$, can be chosen in any direction. Thus the VWE$^{(0)}$ modes correspond to $\mathbf{C}^{\mathbf{S}}_{\mathbf{nv}} \otimes \mathbf{C}^{\mathbf{P}}_{\infty\mathbf{v}}$ symmetry. Furthermore, in determining the geometrically perturbed fiber true vector modes, we note that the problem now involves two competing perturbations: (1) geometrical and (2) polarization (Δ). The form of the modal fields will depend on which has the stronger effect. When polarization splitting is dominant, inclusion of the n-fold geometrical deformation as a small perturbation is given by the $\mathbf{C}^{\mathbf{J}}_{\infty\mathbf{v}} \supset \mathbf{C}^{\mathbf{J}}_{\mathbf{nv}}$ reduction above. When the geometry splitting is dominant, to account for the inclusion of polarization effects upon the VWE$^{(0)}$ modes of the geometrically perturbed fiber we need to consider the reduction $\mathbf{C}^{\mathbf{S}}_{\mathbf{nv}} \otimes \mathbf{C}^{\mathbf{P}}_{\infty\mathbf{v}} \supset \mathbf{C}^{\mathbf{J}}_{\mathbf{nv}}$ for the irreps $l \otimes \mathbf{1}(\mathbf{C}^{\mathbf{S}}_{\mathbf{nv}} \otimes \mathbf{C}^{\mathbf{P}}_{\infty\mathbf{v}})$. For $n > 2$, the irrep $\mathbf{1}(\mathbf{C}^{\mathbf{P}}_{\mathbf{nv}})$ is two-dimensional with the same basis functions as $\mathbf{1}(\mathbf{C}^{\mathbf{P}}_{\infty\mathbf{v}})$, the appropriate reduction is the same as for $\mathbf{C}_{\mathbf{nv}} \otimes \mathbf{C}_{\mathbf{nv}} \supset \mathbf{C}_{\mathbf{nv}}$ for which direct product tables are more readily available (see the Appendix and, e.g., Ref. 10.)

4.1.2 Anticrossing and Mode Form Transitions

Given that we know the order of the modes both in the case when the geometric perturbation dominates and in the case when the polarization splitting dominates, we use the principle [57; 58, vol. 1, pp. 406, 466] that β *levels cannot cross for modes of the same symmetry* to determine the correspondence between the modes for the two cases.

4.2 C_{2v} SYMMETRY: ELLIPTICAL (OR RECTANGULAR) GUIDES: ILLUSTRATION OF METHOD

4.2.1 Wave Equation Symmetries and Mode-Irrep Association

Deformations that retain two axes of symmetry such as those of elliptical or rectangular form correspond to lowering the geometrical symmetry group to $\mathbf{C}_{2\mathbf{v}}$ and hence the SWE symmetry to $\mathbf{C}^{\mathbf{S}}_{2\mathbf{v}}$, the VWE$^{(0)}$ symmetry to $\mathbf{C}^{\mathbf{S}}_{2\mathbf{v}} \otimes \mathbf{C}^{\mathbf{P}}_{\infty\mathbf{v}}$, and the VWE symmetry to $\mathbf{C}^{\mathbf{J}}_{2\mathbf{v}}$.

§ We illustrate the problem with reference to elliptical deformation of a circular fiber. However, the general results also have applicability to rectangular deformations [52, 98, 99] and the dumbbell shape produced by the fusion of two identical fibers [41, 100], etc. §

With regard to the group theoretic apparatus, in Fig. 4.1, the four symmetry operations of the group $\mathbf{C_{2v}}$ are illustrated with respect to an ellipse. In Fig. 4.2 we identify the various ellipse modes for $l = 1$ with the appropriate (one-dimensional) irreps by noting the characters corresponding to symmetry operations of $\mathbf{C_{2v}}$ on the modal fields. The appropriate branching rules $\mathbf{C_{\infty v}} \supset \mathbf{C_{2v}}$ for the resolution of the irreps of $\mathbf{C_{\infty v}}$ into the irreps of $\mathbf{C_{2v}}$ are summarized in Table A.8.

4.2.2 Mode Splittings

Using the above branching rules and mode-irrep associations, in Fig. 4.3 we consider the splitting of modal propagation levels due to an elliptical perturbation.

Ellipticity Splitting Dominates

On the right-hand side of Fig. 4.3, the two $l = 1$ scalar modes transform according to the representation $\mathbf{1}(\mathbf{C}^{S}_{\infty v}) \equiv \mathbf{1}^{S}_{\infty v}$. Inclusion of ellipticity and the corresponding reduction to $\mathbf{C}^{S}_{\infty v} \supset \mathbf{C}^{S}_{2v}$ leads to the splitting of even and odd scalar modes, which transform as $\mathbf{1}^{S}_{2v}$ and $\tilde{\mathbf{1}}^{S}_{2v}$, respectively. In the limit $\Delta = 0$, the inclusion of polarization leads to 2 two-fold degenerate vector mode levels transforming as irreps of the direct product group $\mathbf{C}^{S}_{2v} \otimes \mathbf{C}^{P}_{\infty v}$. The splitting

FIGURE 4.1

Symmetry operations of $\mathbf{C_{2v}}$ on an elliptical cross section.

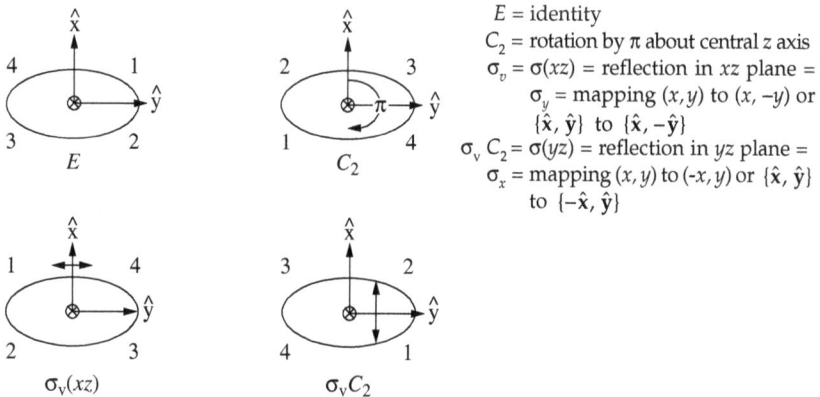

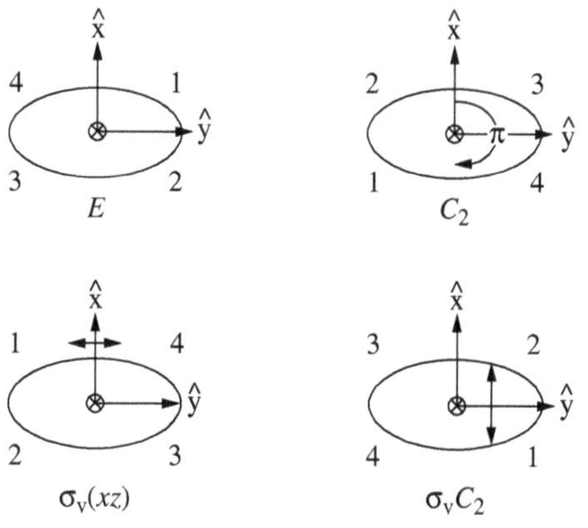

E = identity
C_2 = rotation by π about central z axis
$\sigma_v = \sigma(xz)$ = reflection in xz plane = σ_y = mapping (x, y) to $(x, -y)$ or $\{\hat{\mathbf{x}}, \hat{\mathbf{y}}\}$ to $\{\hat{\mathbf{x}}, -\hat{\mathbf{y}}\}$
$\sigma_v C_2 = \sigma(yz)$ = reflection in yz plane = σ_x = mapping (x, y) to $(-x, y)$ or $\{\hat{\mathbf{x}}, \hat{\mathbf{y}}\}$ to $\{-\hat{\mathbf{x}}, \hat{\mathbf{y}}\}$

FIGURE 4.2

Identification of ellipse modes with C_{2v} irreps: (a) scalar modes, (b) LP modes, and (c) (near) circular fiber vector modes. Symmetry operations on the vector modes are joint (J), i.e., rotation or reflection of the mode pattern as a whole.

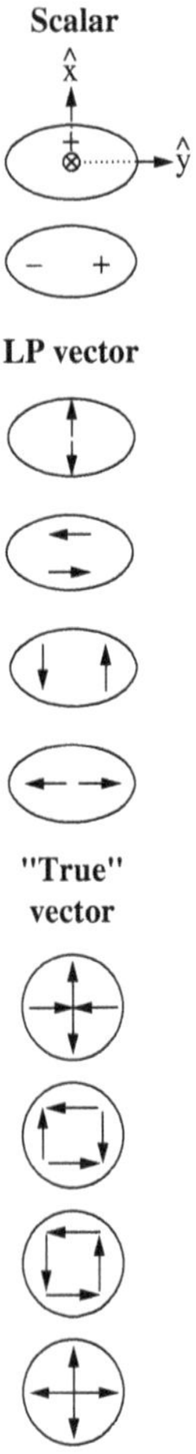

Modes	Symmetry operations				Irreps of	
	E	C_2	σ_v	$\sigma_v C_2$	C_{2v}	
Scalar						
ψ_{11}^e	1	-1	1	-1	**1**	(a)
ψ_{11}^o	1	-1	-1	1	$\tilde{1}$	
LP vector						
LP_{11}^{ex}	1	1	1	1	**0**	
LP_{11}^{ey}	1	1	-1	-1	$\tilde{0}$	(b)
LP_{11}^{ox}	1	1	-1	-1	$\tilde{0}$	
LP_{11}^{oy}	1	1	1	1	**0**	
"True" vector						
HE_{21}^e	1	1	1	1	**0**	
HE_{21}^o	1	1	-1	-1	$\tilde{0}$	(c)
TE_{01}	1	1	-1	-1	$\tilde{0}$	
TM_{01}	1	1	1	1	**0**	

All irreps of C_{2v} are one-dimensional. Thus modal identification with these irreps is simply given by examining the character table

For example:
The even scalar mode has its pattern inverted (-1) by π rotations and by reflections in the yz plane. This corresponds to the irrep $1(C_{2v})$ as in Table A.7.

HE_{2m}^e modes are invariant under all symmetry operations of C_{2v} ⇒ association with the identity irrep $0(C_{2v})$. Cf. HE_{2m}^o modes which are inverted by both reflections ⇒ association with $\tilde{0}(C_{2v})$. (Note: HE_{2m}^o are simply $\pi/4$ rotations of HE_{2m}^e—not all arrows are shown here for the case $m = 1$, i.e., HE.)

resulting from finite Δ is then given by the reduction $C_{2v}^S \otimes C_{\infty v}^P \supset C_{2v}^S$; e.g., for the even modes the corresponding branching rule is $1_{2v}^S \otimes 1_{\infty v}^P \rightarrow 0_{2v}^J \oplus \tilde{0}_{2v}^J$ with the Clebsch-Gordan coefficients confirming the linearly polarized modes LP_{1m}^{ex} and LP_{1m}^{ey} as the basis functions corresponding to 0_{2v}^J and $\tilde{0}_{2v}^J$, respectively, in this even mode case.

FIGURE 4.3

Transformation of $lm = 11$ circular fiber vector modes to ellipse modes for (*a*) $V < 3.8$ and (*b*) $V > 3.8$. On the left-hand side of the diagram, polarization splitting dominates; on the right, the effects of ellipticity dominate. Modes are ordered with $\beta = kn_{\mathrm{eff}}$ increasing toward the top of the page.

§ Symmetry groups (top row) are those of a circle $\mathbf{C}_{\infty v}$, that of an ellipse $\mathbf{C}_{2v}$, and direct products ($\otimes$) thereof. Superscripts S, P, and J refer respectively to scalar, polarization, and joint symmetries. $\supset \tilde{\mathbf{C}}_{nv}$ indicates symmetry reduction with respect to subgroup $\mathbf{C}_{nv}$ with corresponding irrep branching rules (e.g., $\mathbf{1} \otimes \mathbf{1} \rightarrow \mathbf{0} \oplus \tilde{\mathbf{0}} \oplus \mathbf{2}$ for $\mathbf{C}_{\infty v} \otimes \mathbf{C}_{\infty v} \supset \mathbf{C}_{\infty v}$) giving the modal level splitting, and irrep dimension giving the level degeneracy. Irreps of $\mathbf{C}_{\infty v}$ (distinguished by subscript $\infty\mathbf{v}$) are $\mathbf{0}, \tilde{\mathbf{0}}, \mathbf{1}, \mathbf{2}, \ldots$ (i.e., respectively, $\mathbf{A}_1, \mathbf{A}_2, \mathbf{E}_1, \mathbf{E}_2, \ldots$ in Mullikans' notation) and are, respectively, of dimension **1, 1,** and **2** thereafter. Irreps of $\mathbf{C}_{2v}$ (distinguished by subscript **2v**) are $\mathbf{0}, \tilde{\mathbf{0}}, \mathbf{1}, \mathbf{1}$ (i.e., $\mathbf{A}_1, \mathbf{A}_2, \mathbf{B}_1, \mathbf{B}_2$) and are all one-dimensional. (As groups are the same in any column, irrep indices are only included for the top and bottom cases.) §

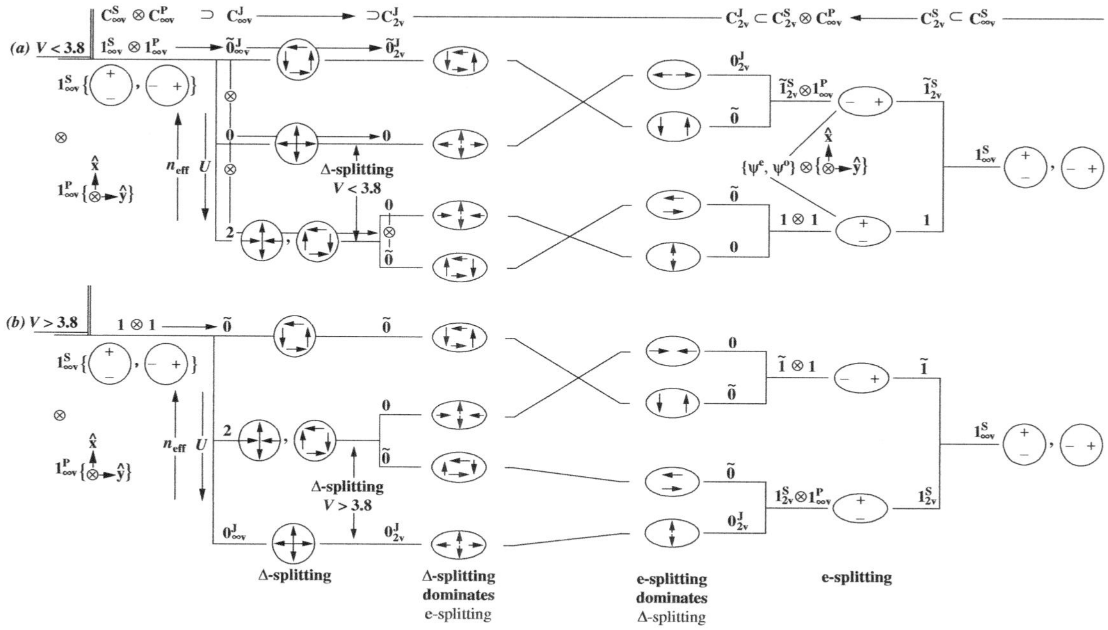

Polarization Splitting Dominates

As we can see from Table A.8, each two-dimensional representation $\mathbf{v}^J_{\infty v} \equiv \mathbf{v}(\mathbf{C}^J_{\infty v}) \geq \mathbf{1}$ is split into two irreps of $\mathbf{C}^J_{2v}$; all HE and EH modes of a circular fiber are also split upon the introduction of ellipticity. For example, in the case $l = 1$ of Fig. 4.3, since the odd and even polarized HE_{2m} modes transform as the representation $\mathbf{2}^J_{\infty v}$, as seen in Sec. 3.3, the "levels" HE_{2m} are each split into two levels with two different β transforming as $\mathbf{0}^J_{2v}$ and $\tilde{\mathbf{0}}^J_{2v}$.

4.2.3 Vector Mode Form Transformations for Competing Perturbations

In Chap. 3, we saw that the circular fiber true modes TM_{0m} and HE^e_{2m} are both constructed from equal combinations of LP^{ex}_{1m} and LP^{oy}_{1m}. Which LP component is favored as the circular guide is increasingly squashed into an ellipse? Both of the true modes TM_{0m} and HE^e_{2m} correspond to $\mathbf{0}^J_{2v}$ when the symmetry is reduced to $\mathbf{C}^J_{2v}$, just as the two LP modes transform as $\mathbf{0}^J_{2v}$. Given that the propagation constants of modes corresponding to the same irrep cannot cross (without interaction) for an adiabatic transformation, the resulting LP mode will depend on whether it is TM_{0m} or HE^e_{2m} that has the larger propagation constant. Given that the order changes [3, p. 899 (Figs. 12-4, 14-5)] at $V \approx 3.8$, we also expect the transformations to change. In particular, for $V > 3.8$, TM_{01} evolves to become LP^{ex}_{11}, that is, the LP mode with the largest propagation constant (smallest U); however, for $V < 3.8$, TM_{01} evolves to become LP^{oy}_{11}. These conclusions are in agreement with the explicit form of the fields determined using perturbation theory and given in Table 13-1 of Ref. 3 (p. 288).

The transformation of higher-order $HE_{\nu m}$ as well as $EH_{\nu m}$ modes to LP ellipse modes is similarly deduced.

4.3 C_{3v} SYMMETRY: EQUILATERAL TRIANGULAR DEFORMATIONS

Examples of lowering the symmetry to $\mathbf{C}_{3v}$ via equilateral triangular deformation are given in Fig. 4.4.

For $\mathbf{C}_{3v}$ there are 2 one-dimensional irreps $\mathbf{0}$ and $\tilde{\mathbf{0}}$, and 1 two-dimensional irrep $\mathbf{1}$ (that is, $\mathbf{A}_1$, $\mathbf{A}_2$, and $\mathbf{E}$, respectively, in Mullikan's notation—see Table A.5 with $n = 3$). The physical significance of the two-dimensional irrep is that in contrast to $\mathbf{C}_{2v}$ deformations which split all modes, equilateral triangular perturbations leave all

FIGURE 4.4

Increasing (equilateral) triangular "perturbations" (t) of a circular fiber: (*a*) by a triangular squashing/deformation and (*b*) by placing three equally spaced cladding depressions/rises and increasing their proximity to the core, or their depth/height. Analogous schemes can be arranged for higher-order deformations.

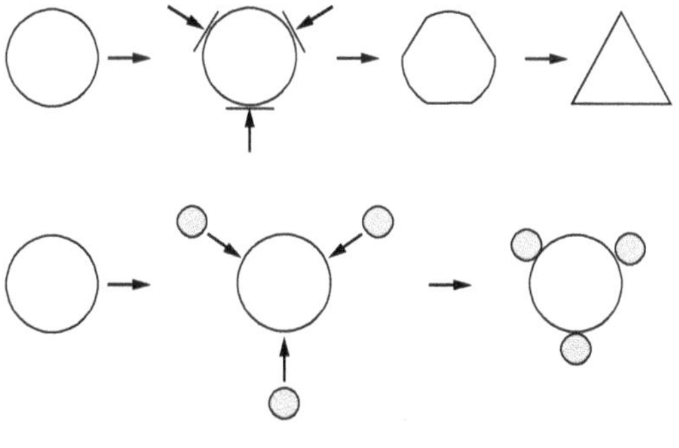

even and odd *scalar modes* Ψ^{e}_{lm} and Ψ^{0}_{lm} degenerate **except those with l a multiple of 3 which are split** by the $\mathbf{C_{3v}}$ deformation. Analogously, for *vector modes*, degeneracy remains between all even and odd polarized pairs of modes $HE^{e}_{\nu m}$ and $HE^{o}_{\nu m}$ (and between $EH^{e}_{\nu m}$ and $EH^{0}_{\nu m}$) **except those with** ν $(= l \pm 1)$ **a multiple of 3 which are split** by the deformation. An example of this is seen in Fig. 4.5, which was generated analogously to the elliptical fiber results of Sec. 4.2 and shows the effect of (1) triangular perturbation being dominant on the right and (2) polarization (finite Δ) dominant on the left. Note also that the *order of the modes*, and thus their *evolution*, in some cases depends on the form of the triangular perturbation. For example, for three *cladding depressions* $\beta(LP^{e}_{3m}) > \beta(LP^{o}_{3m})$ and thus as $\beta(HE_{4m}) > \beta(EH_{2m})$, the principle of anticrossing of levels corresponding to the same representation means that the HE_{4m} pairs evolve to LP^{e}_{3m} pairs and the EH_{2m} pairs to the LP^{o}_{3m} pairs. For raised cladding perturbations, the opposite transitions occur.

§ In the limit of an equilateral triangle, when the field is no longer simply a small perturbation of that of a circular fiber, construction of the scalar mode field *in the limit of a step profile with large V* (i.e., for $\psi = 0$ on the triangle boundary) has been carried out for the equivalent two-dimensional Schrödinger equation in Ref. 101 (p. 1085) using projection operator methods (i.e., the methods used here in Chap. 5 for multicore fibers—see also Ref. 91, p. 237, for an intuitive textbook discussion of symmetry consequences for construction of some of the modes). Given the scalar modes and thus the LP pseudo-modes,

FIGURE 4.5

Transformation of circular fiber vector modes to triangularly perturbed fiber modes. On the left-hand side of the diagram, polarization splitting dominates; on the right, the effects of triangularity dominate.

§ Irreps of $\mathbf{C_{3v}}$ are $\mathbf{0}$, $\tilde{\mathbf{0}}$, $\mathbf{1}$ (i.e., $\mathbf{A_1}$, $\mathbf{A_2}$, $\mathbf{E}$) of dimension **1**, **1**, and **2**. §

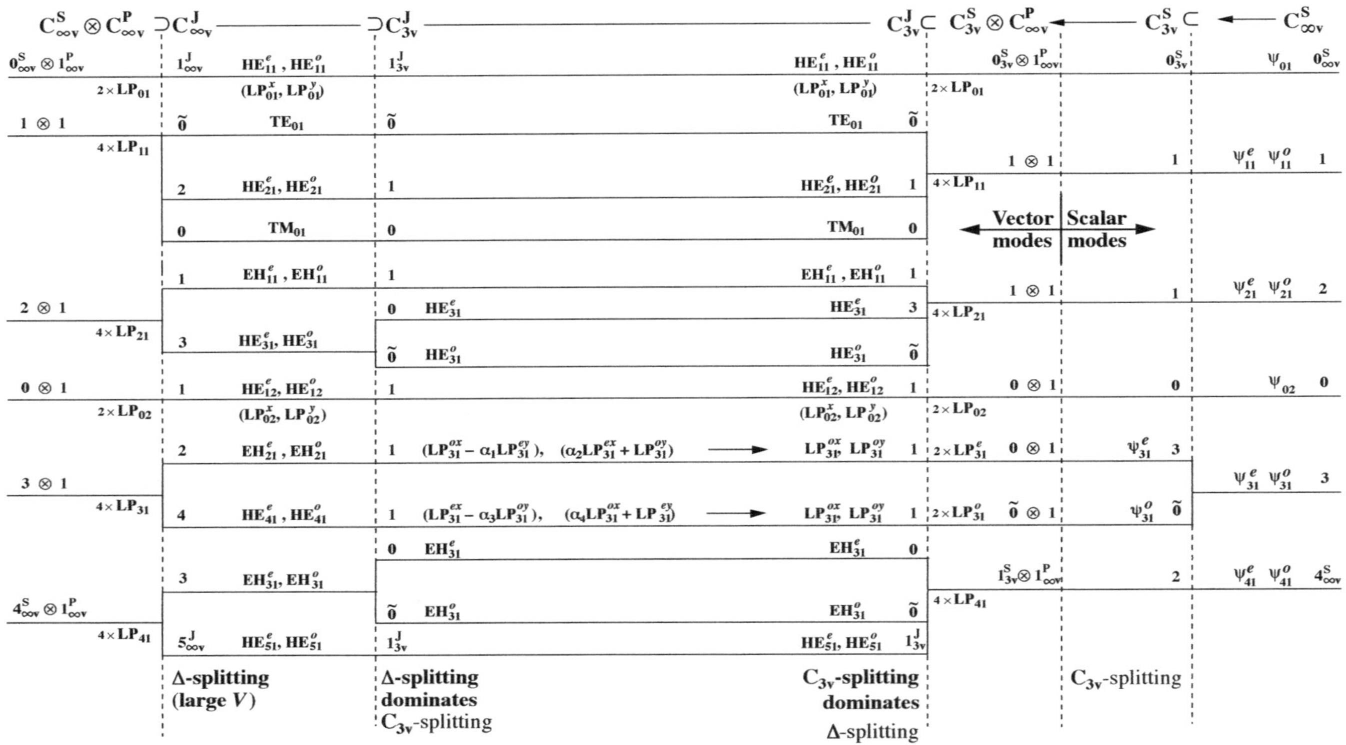

the true weak-guidance vector fields are then given as the same combinations as in Fig. 4.5 [i.e., those obtained using the Clebsch-Gordan (CG) coefficients (Sec. A.4) for the symmetry reduction $\mathbf{C}_{3v} \otimes \mathbf{C}_{\infty v} \supset \mathbf{C}_{3v}$]. §

4.4 C_{4V} SYMMETRY: SQUARE DEFORMATIONS

4.4.1 Irreps and Branching Rules

$\mathbf{C}_{4v}$ has irreps $\mathbf{0}, \tilde{\mathbf{0}}, \mathbf{2}, \tilde{\mathbf{2}}$ which are one-dimensional and **1** which is two-dimensional. (These correspond to $\mathbf{A}_1$, $\mathbf{A}_2$, $\mathbf{B}_1$, $\mathbf{B}_2$, and **E**, respectively, in Mullikan's notation—see Table A.5 with $n = 4$.)

$\mathbf{C}_{\infty v} \supset \mathbf{C}_{4v}$: From Ref. 9 (p. 213) we have the branching rules as $\mathbf{0} \to \mathbf{0}$, $\tilde{\mathbf{0}} \to \tilde{\mathbf{0}}$, all odd irreps branch to **1**, and even irreps split as $(\mathbf{4n}-\mathbf{2}) \to \mathbf{2} \oplus \tilde{\mathbf{2}}$ and $\mathbf{4n} \to \mathbf{0} \oplus \tilde{\mathbf{0}}$ for $n \geq 1$.

$\mathbf{C}_{4v} \otimes \mathbf{C}_{\infty v} \supset \mathbf{C}_{4v}$: Noting the equivalence of the direct product reductions $p \otimes \mathbf{1}(\mathbf{C}_{4v} \otimes \mathbf{C}_{\infty v})$ with those of the subgroup $\mathbf{C}_{4v} \otimes \mathbf{C}_{4v}$, from Ref. 10 (p. 14, Table 4) we have (1) $\mathbf{1} \otimes \mathbf{1} \to \mathbf{0} \oplus \tilde{\mathbf{0}} \oplus \mathbf{2} \oplus \tilde{\mathbf{2}}$ and (2) $p \otimes \mathbf{1} \to \mathbf{1}$ for $p = \mathbf{0}, \tilde{\mathbf{0}}, \mathbf{2}$, and $\tilde{\mathbf{2}}$.

4.4.2 Mode Splitting and Transition Consequences

For a $\mathbf{C}_{4v}$ perturbation of a circular fiber, mode splittings and transitions are shown in Fig. 4.6. for the case of symmetric raised perturbations centered on symmetry axes located at angles $\phi = n\pi/2$ or depressed perturbations between these, i.e., at angles $\phi = (2n + 1)\pi/4$.

For the *scalar modes* (corresponding to the first reduction in the form $\mathbf{C}^{\mathrm{S}}_{\infty v} \supset \mathbf{C}^{\mathrm{S}}_{4v}$), **splitting occurs between even and odd modes** Ψ^{e}_{lm} and Ψ^{o}_{lm} **for** l **even**. Similarly, **for** *vector modes* (corresponding to $\mathbf{C}^{\mathrm{J}}_{\infty v} \supset \mathbf{C}^{\mathrm{J}}_{4v}$), **splitting occurs between the even and odd polarizations** of $\mathrm{HE}_{\nu m}$ and $\mathrm{EH}_{\nu m}$ **for** ν **even** (that is, $l = \nu \pm 1$ odd). Scalar modes with l odd and vector modes with ν odd remain degenerate.

The direct product reduction (1) corresponds to a **four-fold Δ-splitting of the LP_{lm} mode levels with odd l** [as the corresponding scalar modes transform as the irrep $\mathbf{1}(\mathbf{C}^{\mathrm{S}}_{4v})$]. Furthermore, it may be shown that the appropriate CG coefficients lead to the **odd l** split

FIGURE 4.6

Transformation of circular fiber vector modes to $\mathbf{C_{4v}}$ perturbed fiber modes. On the left-hand side of the diagram, polarization splitting dominates; on the right, the effects of $\mathbf{C_{4v}}$ splitting dominate.

§ Irreps of $\mathbf{C_{4v}}$ are **0**, $\tilde{\mathbf{0}}$, **1**, **2**, $\tilde{\mathbf{2}}$ (i.e., $\mathbf{A_1}$, $\mathbf{A_2}$, **E**, $\mathbf{B_1}$, $\mathbf{B_2}$) of dimension **1**, **1**, **2**, **1**, and **1**. §

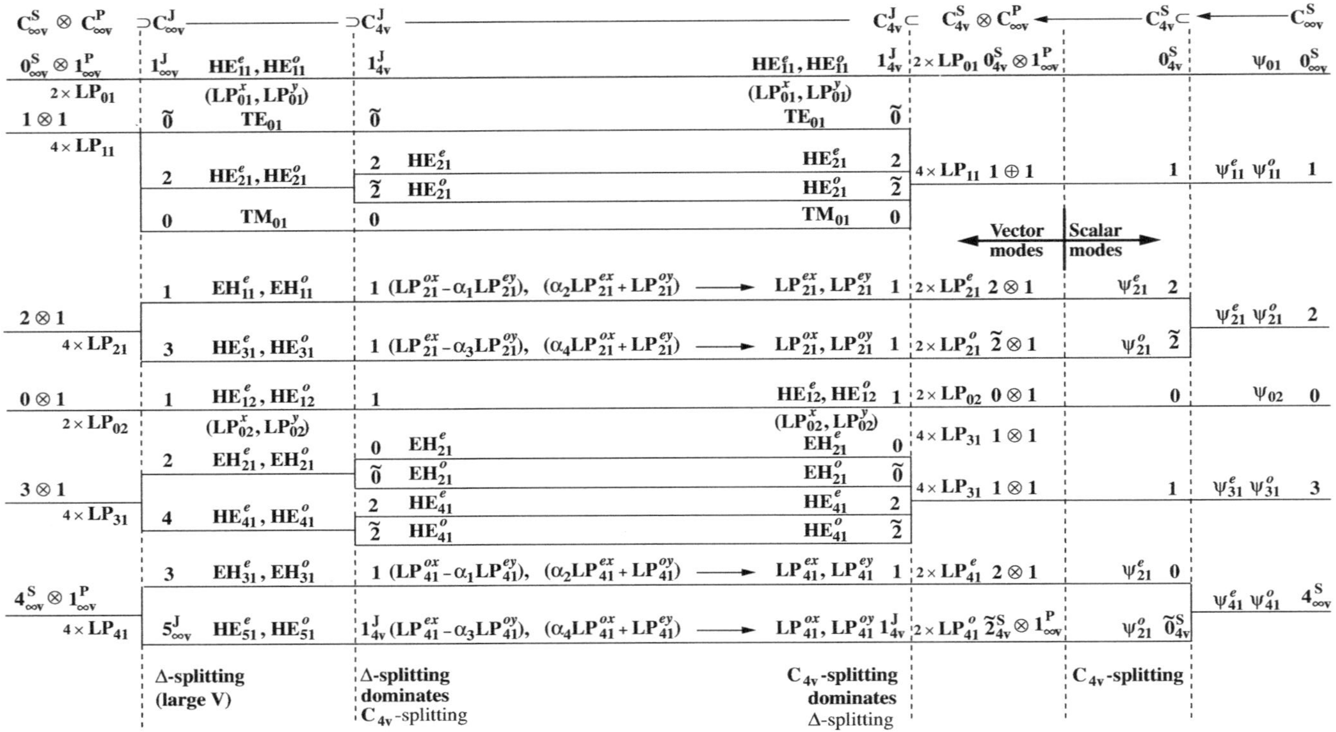

levels being the same LP_{lm} as for the circular case, **but with all degeneracy removed;** e.g., LP_{1m} splits into TM_{0m}, TE_{0m}, and nondegenerate even and odd HE_{2m} modes. Note that this is in contrast to the triangular case ($n = 3$) for which the HE_{2m} modes remain degenerate. The direct product reduction (2) corresponds to **degeneracy remaining between pairs of LP modes with the same polarization**, that is, LP^{ex}_{lm} and LP^{ox}_{lm} (or LP^{ey}_{lm} and LP^{oy}_{lm}), **if l is even.** However, note that these LP pairs, being of the same polarization, are the modes that must undergo a form transition as we increase the relative strength of Δ with respect to the square perturbation; e.g., the two LP_{2m} levels will make transitions to the HE_{3m} and EH_{1m} levels in a manner analogous to that shown in Fig. 4.5 for LP_{31} in the triangular case. Again, whether it is the even or odd LP_{2m} level that makes the transition to HE_{3m} and vice versa for EH_{1m} depends on the details of the perturbations.

4.4.3 Square Fiber Modes and Extra Degeneracies

§ In the limit deformation from a circle to a square with large V, extra degeneracy may occur that is more than that given by $\mathbf{C}_{4v}$ symmetry as discussed in Ref. 102 in the quantum mechanical context. While "square"-symmetric deformation of a step profile results in the splitting of some circular fiber modes as described by the $\mathbf{C}_{\infty v} \supset \mathbf{C}_{4v}$ reduction, in the large V limit (i.e., zero field boundary condition) as the fiber takes on a square shape, some modes tend to coalesce. For example, while the even and odd ψ_{21} modes split, one of the ψ_{21} modes tends toward degeneracy with ψ_{02}. In the absolute square limit these are simply the combinations $E_{31} \pm E_{13}$, where $E_{pq} = \sin(U_p X) \sin(U_q Y)$ with $U_p = p\pi/2$, etc., which result from the obvious degeneracy of E_{mn} and E_{nm}. This extra degeneracy related to separability in X and Y (in the large V limit) is similar to that resulting from the "hidden symmetry" in the case of a parabolic profile circular fiber mentioned in Sec. 3.2.2—see also Ref. 1. §

4.5 C_{5V} SYMMETRY: PENTAGONAL DEFORMATIONS

4.5.1 Irreps and Branching Rules

C_{5v} has irreps **0** and $\tilde{\mathbf{0}}$ which are one-dimensional and **1** and **2** which are two-dimensional. (These correspond to $\mathbf{A}_1$, $\mathbf{A}_2$, $\mathbf{E}_1$, and $\mathbf{E}_2$, respectively, in Mullikan's notation—see Table A.5 with $n = 5$.)

$\mathbf{C}_{\infty v} \supset \mathbf{C}_{5v}$: From Ref. 9 (p. 213) we have the branching rules as $\mathbf{0} \to \mathbf{0}$, $\tilde{\mathbf{0}} \to \tilde{\mathbf{0}}, \mathbf{1} \to \mathbf{1}, \mathbf{2} \to \mathbf{2}$; and odd irreps branch to **1** and even irreps split as $(\mathbf{5}n \pm \mathbf{2}) \to \mathbf{2}$, $(\mathbf{5}n \pm \mathbf{1}) \to \mathbf{1}$, and $\mathbf{5}n \to \mathbf{0} \oplus \tilde{\mathbf{0}}$ for $n \geq 1$. $\mathbf{C}_{5v} \otimes \mathbf{C}_{\infty v} \supset \mathbf{C}_{5v}$: Noting the equivalence of the direct product reductions $p \otimes \mathbf{1}(\mathbf{C}_{5v} \otimes \mathbf{C}_{\infty v})$ with those of the subgroup $\mathbf{C}_{5v} \otimes \mathbf{C}_{5v}$, from Ref. 10 (p. 14, Table 4) we have (1) $\mathbf{1} \otimes \mathbf{1} \to \mathbf{0} \oplus \tilde{\mathbf{0}} \oplus \mathbf{2}$, (2) $\mathbf{2} \otimes \mathbf{1} \to \mathbf{1} \oplus \mathbf{2}$, and (3) $p \otimes \mathbf{1} \to \mathbf{1}$ for $p = \mathbf{0}$ and $\tilde{\mathbf{0}}$.

4.5.2 Mode Splitting and Transition Consequences

For a $\mathbf{C}_{5v}$ perturbation of a circular fiber, mode splittings and transitions are shown in Fig. 4.7 for the case of symmetric raised perturbations centered on symmetry axes located at angles $\phi = 2n\pi/5$ or depressed perturbations between these, i.e., at angles $\phi = (2n + 1)\pi/5$.

For the *scalar modes* (corresponding to the first reduction in the form $\mathbf{C}^{\mathrm{S}}_{\infty v} \supset \mathbf{C}^{\mathrm{S}}_{5v}$, **splitting occurs between even and odd modes** Ψ^{e}_{lm} and Ψ^{o}_{lm} **for l a multiple of 5**. Similarly, **for** *vector modes* (corresponding to $\mathbf{C}^{\mathrm{J}}_{\infty v} \supset \mathbf{C}^{\mathrm{J}}_{5v}$), **splitting occurs between the even and odd polarizations** of $\mathrm{HE}_{\nu m}$ and $\mathrm{EH}_{\nu m}$ **for ν a nonzero multiple of 5.** Scalar modes with l not a multiple of 5 and vector modes with ν not a multiple of 5 remain degenerate.

The direct product reduction (1) corresponds to a **three-fold Δ-splitting of the LP_{lm} mode levels for $l = 1$ and 4** [as the corresponding scalar modes transform as the irrep $\mathbf{1}(\mathbf{C}^{\mathrm{S}}_{5v})$]. The direct product reduction (2) corresponds to **degeneracy remaining between pairs of LP modes with the same polarization**, that is, LP^{ex}_{lm} and LP^{ox}_{lm} (or LP^{ey}_{lm} and LP^{oy}_{lm}), **if l is a nonzero multiple of 5.** However, note that these LP pairs, being of the same polarization, are the modes that must undergo a form transition as we increase the relative strength of Δ with respect to the pentagonal perturbation; e.g., the two LP_{5m} levels will make transitions to the HE_{6m} and EH_{4m} levels in a manner analogous to that shown in Fig. 4.5 for LP_{31} in the triangular case. Again, whether it is the even or odd LP_{5m} level that makes the transition to HE_{6m} and vice versa for EH_{4m} depends on the details of the perturbations.

FIGURE 4.7

Transformation of circular fiber vector modes to pentagonally perturbed fiber modes. On the left-hand side of the diagram, polarization splitting dominates; on the right, the effects of pentagonality dominate.

Irreps of $\mathbf{C_{5v}}$ are **0**, $\tilde{\mathbf{0}}$, **1**, **2** (i.e., $\mathbf{A_1}$, $\mathbf{A_2}$, $\mathbf{E_1}$, $\mathbf{E_2}$) of dimension **1**, **1**, **2**, and **2**, respectively.

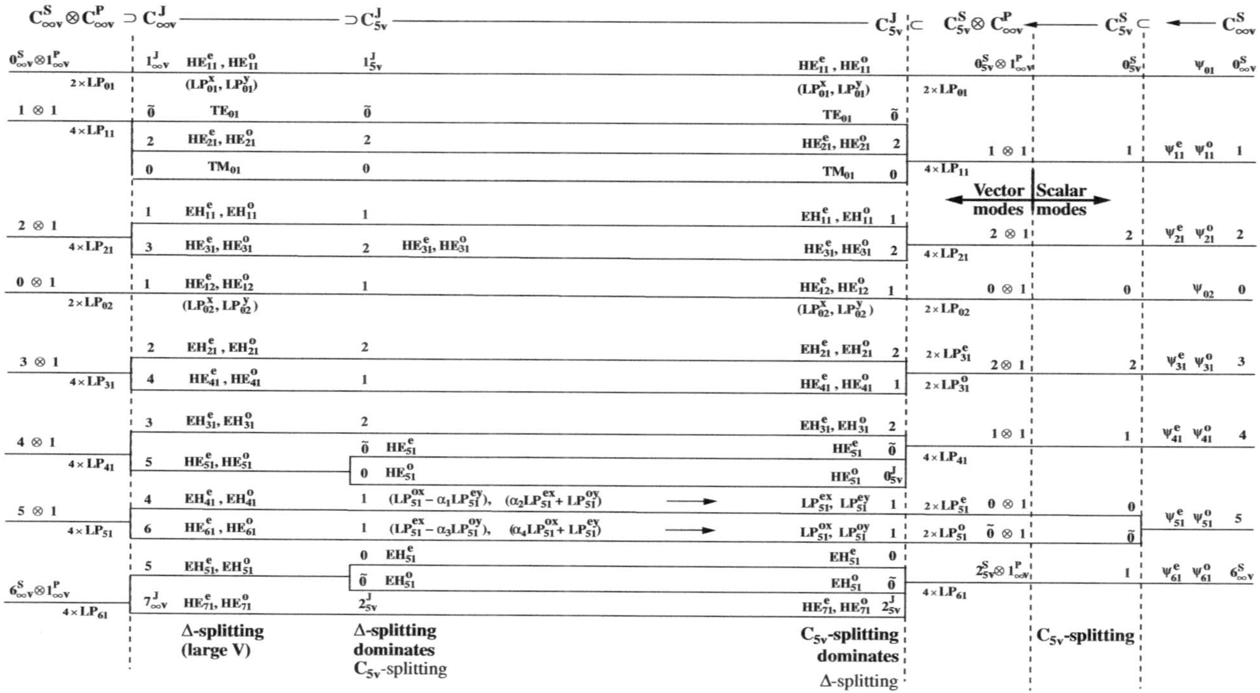

4.6 C_{6v} SYMMETRY: HEXAGONAL DEFORMATIONS

4.6.1 Irreps and Branching Rules

C_{6v} has irreps $\mathbf{0}, \tilde{\mathbf{0}}, \mathbf{3}$, and $\tilde{\mathbf{3}}$ which are one-dimensional and **1** and **2** which are two-dimensional. (These correspond to $\mathbf{A}_1$, $\mathbf{A}_2$, $\mathbf{B}_1$, and $\mathbf{B}_2$, and $\mathbf{E}_1$ and $\mathbf{E}_2$, respectively, in Mullikan's notation—see Table A.5 with $n = 6$.)

$C_{\infty v} \supset C_{6v}$: From Ref. 9 (p. 213) we have the branching rules as $\mathbf{0} \to \mathbf{0}$, $\tilde{\mathbf{0}} \to \tilde{\mathbf{0}}, \mathbf{1} \to \mathbf{1}, \mathbf{2} \to \mathbf{2}$, and for $n \geq 1, \mathbf{6n \pm 3} \to \mathbf{3} \oplus \mathbf{3}$, $\mathbf{6n \pm 2} \to \mathbf{2}, \mathbf{6n \pm 1} \to \mathbf{1}, \mathbf{6n} \to \mathbf{0} \oplus \tilde{\mathbf{0}}$.

$C_{6v} \otimes C_{\infty v} \supset C_{6v}$: Noting the equivalence of the direct product reductions $p \otimes \mathbf{1}(C_{6v} \otimes C_{\infty v})$ with those of the subgroup $C_{6v} \otimes C_{6v}$, from Ref. 10 (p. 14, Table 4) we have (1) $p \otimes \mathbf{1} \to \mathbf{1}$ for $p = \mathbf{0}, \tilde{\mathbf{0}}$; (2) $p \otimes \mathbf{1} \to \mathbf{2}$ for $p = \mathbf{3}$ and $\tilde{\mathbf{3}}$; (3) $\mathbf{1} \otimes \mathbf{1} \to \mathbf{0} \oplus \tilde{\mathbf{0}} \oplus \mathbf{2}$; and (4) $\mathbf{2} \otimes \mathbf{1} \to \mathbf{0} \oplus \mathbf{3} \oplus \tilde{\mathbf{3}}$.

Although the number of irreps is doubled in comparison with simple triangular symmetry, the reduction $C_{6v} \supset C_{3v}$ shows no further splitting; i.e., the two two-dimensional irreps $\mathbf{1}(C_{6v})$ and $\mathbf{2}(C_{6v})$ both transform to the one two-dimensional irrep $\mathbf{1}(C_{3v})$. [For the one-dimensional irreps, $\mathbf{0}(C_{6v})$ and $\mathbf{3}(C_{6v}) \to \mathbf{0}(C_{3v})$ and $\tilde{\mathbf{0}}(C_{6v})$ and $\tilde{\mathbf{3}}(C_{6v}) \to \tilde{\mathbf{0}}(C_{3v})$].

4.6.2 Mode Splitting and Transition Consequences

For a C_{6v} perturbation of a circular fiber, the degeneracy breaking is the same as for C_{3v} symmetry. The mode splittings and transitions are shown in Fig. 4.8 for the case of symmetric raised perturbations centered on symmetry axes located at angles $\phi = n\pi/3$ or depressed perturbations between these, i.e., at angles $\phi = (2n + 1)\pi/6$.

For the *scalar modes* (corresponding to the first reduction in the form $C^S_{\infty v} \supset C^S_{6v}$), **splitting occurs between even and odd modes** Ψ^e_{lm} and Ψ^o_{lm} for **nonzero l a multiple of 3.**

Similarly, **for** *vector modes* (corresponding to $C^J_{\infty v} \supset C^J_{6v}$), **splitting occurs between the even and odd polarizations** of $HE_{\nu m}$ and $EH_{\nu m}$ **for nonzero ν being a multiple of 3.** The direct product reduction (1) and (2) corresponds to **degeneracy remaining between pairs of LP modes with the same polarization,** that is, LP^{ex}_{lm} and LP^{ox}_{lm} (or LP^{ey}_{lm} and LP^{oy}_{lm}), **if l is a multiple of 3.**

FIGURE 4.8

Transformation of circular fiber vector modes to triangularly perturbed fiber modes. On the left-hand side of the diagram, polarization splitting dominates; on the right, the effects of triangularity dominate. **Irreps** of $\mathbf{C_{6v}}$ are $\mathbf{0}$, $\mathbf{\tilde{0}}$, $\mathbf{1}$, $\mathbf{2}$, $\mathbf{3}$, $\mathbf{\tilde{3}}$ (i.e., $\mathbf{A_1}$, $\mathbf{A_2}$, $\mathbf{E_1}$, $\mathbf{E_2}$, $\mathbf{B_1}$, $\mathbf{B_2}$) of dimension 1, 1, 2, 2, 1, 1, respectively.

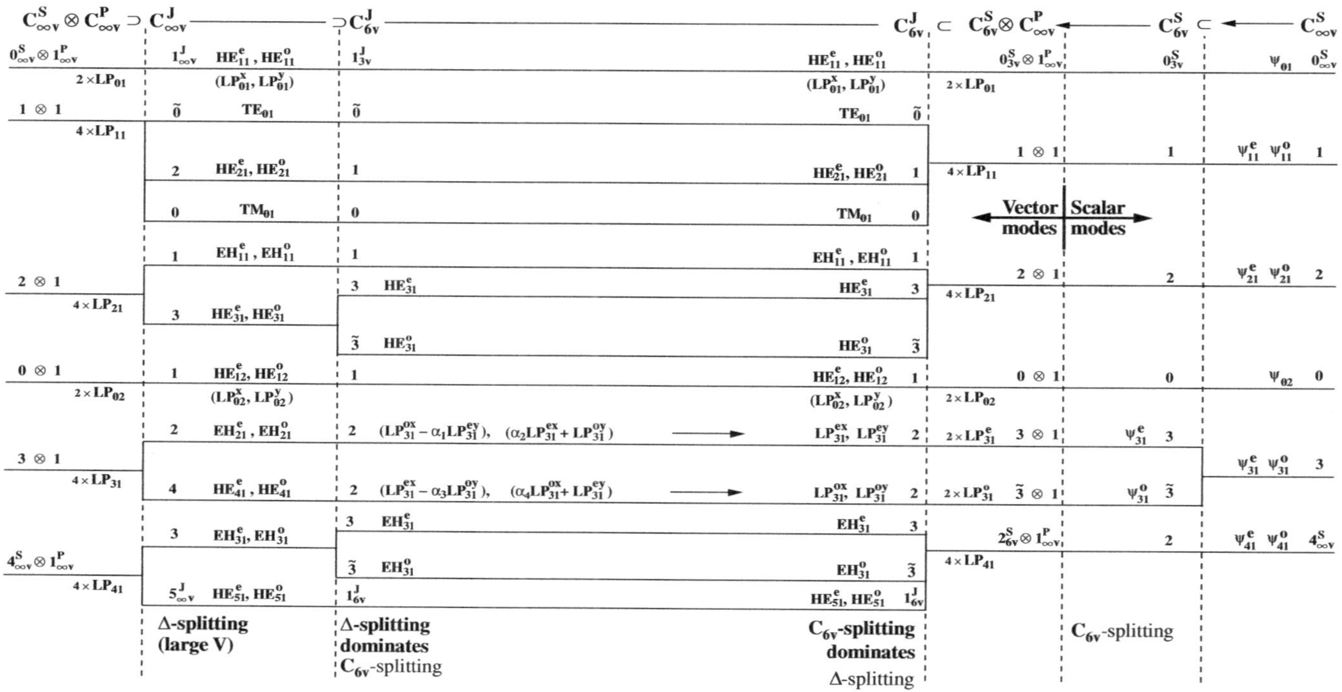

However, note that these LP pairs, being of the same polarization, are the modes that must undergo a form transition as we increase the relative strength of Δ with respect to the square perturbation; e.g., the two LP_{3m} levels will make transitions to the HE_{4m} and EH_{2m} levels. Whether it is the even or odd LP_{2m} level that makes the transition to HE_{4m} and vice versa for EH_{2m} depends on the details of the perturbations.

4.7 LEVEL SPLITTING QUANTIFICATION AND FIELD CORRECTIONS

§ Quantification of the scalar mode propagation constant shifts (or "corrections") and splittings due to geometrical perturbation is straightforwardly undertaken using the reciprocity relation of Eq. (2.28) with the "barred" quantities referring to the circular (unperturbed) fiber and the unbarred quantities to the geometrically perturbed fiber together with the approximation to first order in the geometrical perturbation of $\psi = \psi$, i.e., replacing the perturbed field by the unperturbed field.

For example, see Ref. 3 (Sec. 18.10) for elliptical fiber modes of azimuthal orders $l = 0$ and 1, and note that if the unperturbed circular fiber is chosen to have the same area as the ellipse, then the fundamental mode propagation constants are identical, i.e., the correction is zero. This is a particularly useful result of more general applicability as discussed and exploited in Ref. 52.

For the vector mode correction splittings, some care is needed in that we now have two competing perturbations and must consider the perturbed modal field due to one of the perturbations before obtaining the propagation constant corrections due to both perturbations. For example, for the fundamental mode of an elliptical fiber, in Ref. 3 (Sec. 18.10) the first-order correction in ellipticity to the scalar field ψ is obtained and then used to derive the polarization splitting of the two fundamental vector modes.

Group theoretically, the symmetry-related simplifications in the evaluation of the above propagation constant and field corrections may be formally obtained using the Wigner-Eckart theorem (see Sec. A.1.2 and cf. Sec. 3.4). Such simplifications are particularly valuable for the higher-order geometrical perturbations when polarization effects are also considered. §

CHAPTER 5

Birefringence: Linear, Radial, and Circular

In this chapter we develop the wave equation of Sec. 2.4.3 for *diagonally* anisotropic media in the particular cases of guides with linearly, radially, and circularly birefringent refractive indices, extending to these cases the "weak guidance" type of approach given in Sec. 2.7 for isotropic light guides. For each type of anisotropy we consider the modal transitions depending on the relative strengths of the profile height and birefringence splittings.

5.1 LINEAR BIREFRINGENCE

Consider the introduction of linear birefringence such that the principal axes are aligned with the fiber axes $\hat{\mathbf{x}}$, $\hat{\mathbf{y}}$, $\hat{\mathbf{z}}$ so that $\mathbb{n}^2\mathbf{E} = n_x^2 E_x \hat{\mathbf{x}} + n_y^2 E_y \hat{\mathbf{y}} + n_z^2 E_z \hat{\mathbf{z}}$; i.e., the x-polarized electric field sees a refractive index n_x, etc., such as for the idealized profile of Fig. 5.1.

5.1.1 Wave Equations: Longitudinal Invariance

From the general vector wave equation for diagonally anisotropic media of Eq. (2.11), we have the VWE for the transverse field components of a longitudinally invariant linearly birefringent fiber, which may be written as the two coupled equations

$$\{\nabla_t^2 + k^2 n_x^2 - \beta^2\} e_x = P_{xx} e_x + P_{xy} e_y \tag{5.1a}$$

$$\{\nabla_t^2 + k^2 n_y^2 - \beta^2\} e_y = P_{yx} e_x + P_{yy} e_y \tag{5.1b}$$

FIGURE 5.1

Idealized fiber with equal and uniform core and cladding linear birefringence $\delta_{xy} = \frac{1}{2}\{1 - n_y^2/n_x^2\}$.

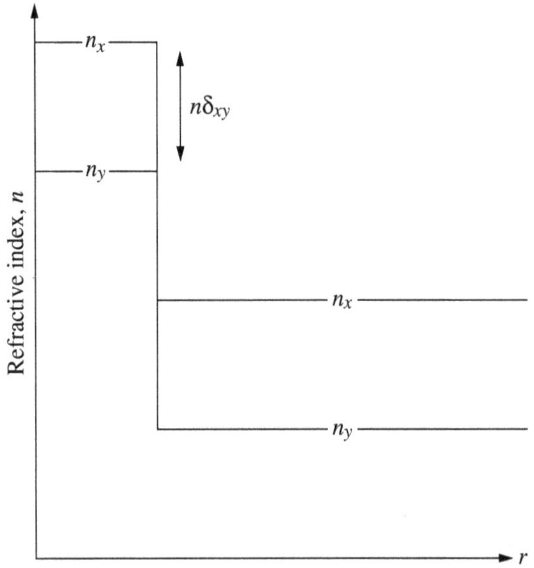

with

$$P_{ij}\, e_j = \partial_i\{2\delta_{zj}\, \partial_j e_j - [1 - 2\delta_{zj}]\, e_j \partial_j (\ln n_j^2)\} \quad \text{and} \quad ij = xx, xy, yx, \text{ or } yy$$

$$\partial_x \equiv \frac{\partial}{\partial x},\ 2\delta_{ij} = 1 - \frac{n_j^2}{n_i^2},\ \text{etc.} \tag{5.1c}$$

Weak Birefringence, Weak Guidance

For weak birefringence with $\delta_{ij} << 1$, in the limit $\Delta \rightarrow 0$, we neglect the right-hand side of Eq. (5.1*a*). Thus each polarization field component is uncoupled, and linearly polarized modes are supported. The x- and y-polarized modes see separate profiles $n_x(x, y)$ and $n_y(x, y)$ with their field magnitudes obeying the scalar wave equations

$$\{\nabla_t^2 + k^2 n_x^2 - \beta^2\} e_x = 0 \tag{5.2a}$$

$$\{\nabla_t^2 + k^2 n_y^2 - \beta^2\} e_y = 0 \tag{5.2b}$$

Stronger Birefringence, Ellipse Analogy, and Even/Odd Mode Splitting

If we continue to consider the limit $\Delta \to 0$, but account for $n_x \neq n_z \neq n_y$ in the equations for the magnitude of the field components, then

$$\left\{\left(\frac{n_x^2}{n_z^2}\right)\frac{\partial}{\partial x^2} + \frac{\partial}{\partial y^2} + k^2 n_x^2 - \beta^2\right\} e_x = 0 \tag{5.3a}$$

$$\left\{\frac{\partial}{\partial x^2} + \left(\frac{n_y^2}{n_z^2}\right)\frac{\partial}{\partial y^2} + k^2 n_y^2 - \beta^2\right\} e_y = 0 \tag{5.3b}$$

§ ***Elliptical Transformation and Splitting Order*** Considering Eq. (5.3) for large δ_{xy} within the weak guidance limit, while e_x and e_y remain uncoupled and thus the modes remain exactly LP, we see that by making the transformation $X = xn_z/n_x$, $Y = y$, Eq. (5.3*a*) becomes the scalar wave equation for the field of an elliptical guide with ratio of ellipse axes X/Y being n_z/n_x. Thus for $n_z < n_x$, the even mode e_{xe} will see the shorter x axis and thus have a smaller n_{eff} than the odd mode e_{xo}, and, as n_{eff} for the x-polarized modes is bounded by n_{co}^x and $n_{cl}^x \approx n_{co}^x(1-\Delta)$, the splitting in n_{eff} will be of order $n\Delta\delta_{xy}$. However, this will remove all remaining degeneracies. If we also have $n_y < n_z$ ($< n_x$), then $n_{\text{eff}}^{ye} < n_{\text{eff}}^{yo}$, as e/o denotes the field distribution with respect to the x axis. §

5.1.2 Mode Transitions: Circular Symmetry

If we consider circular fibers, then each of the polarization field components e_x, e_y (and e_z) sees a circularly symmetric profile as in Fig. 5.1. The mode splittings dependent on the birefringence and profile height are given in Fig. 5.2. We now discuss these in terms of symmetry.

Weak Guidance Limit for Weak Birefringence

For n_x and n_y satisfying $\mathbf{C}_{\infty v}$ symmetry, within the weak guidance limit $\Delta \to 0$, for small δ_{xy}, the scalar magnitudes of the x- and y-polarized fields as described by the uncoupled Eqs. (5.2) are *separately* governed by $\mathbf{C}_{\infty v}^{\mathbf{S}}$ symmetry. Thus, *if only Eqs. (5.2) are considered* [i.e., we set all $P_{ij}\, e_j = 0$ in Eqs. (5.1)], the scalar magnitudes may have an arbitrary orientation with respect to the birefringent axes.

FIGURE 5.2

Schematic of example of *level splitting due to linear birefringence* $\delta_{xy} = \delta_{xy}(r) > 0$ compared with profile height (Δ) splitting for an idealized circularly symmetric profile such as in Fig. 5.1. The order of the split modes corresponds to $n_x > n_z > n_y$

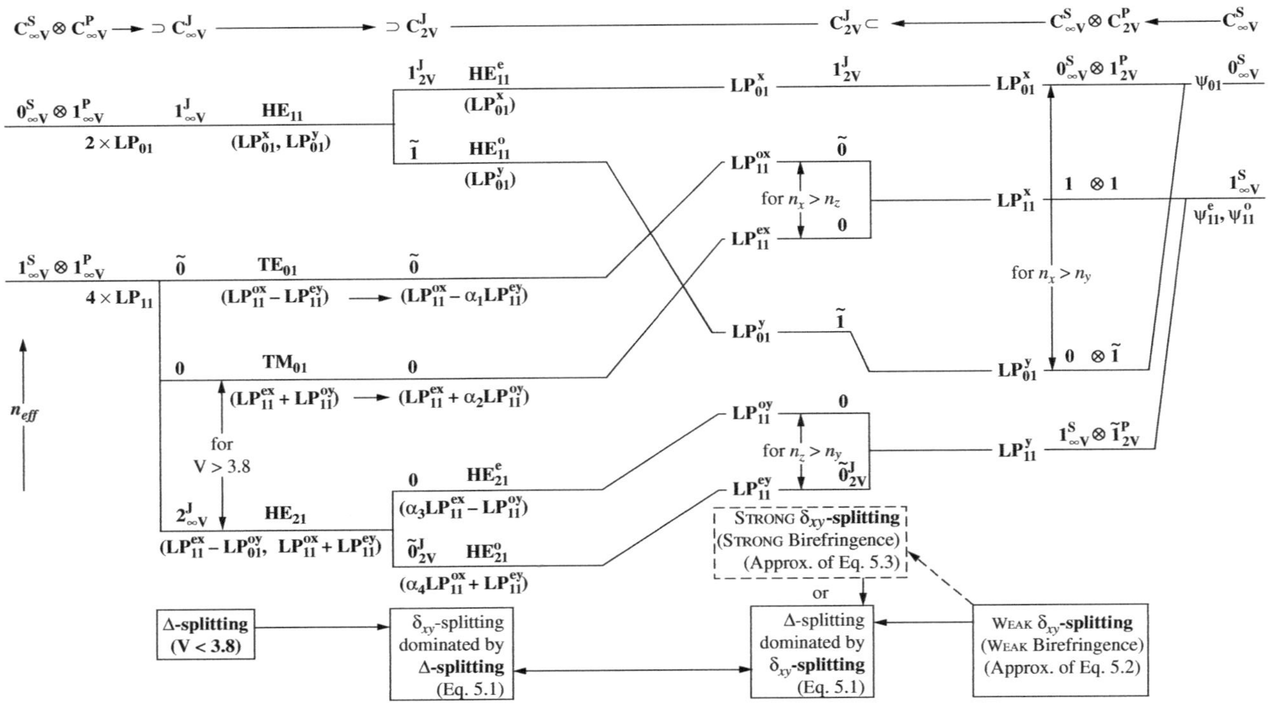

For simplicity, we choose modes to be even and odd with respect to cartesian axes aligned with the birefringent axes, but pairs e_{xe} and e_{xo} etc. remain degenerate.

When the vector nature of the fields is considered, their transverse polarization directions $\hat{\mathbf{x}}$ and $\hat{\mathbf{y}}$ are restricted to the two birefringent axes. They are governed by $\mathbf{C}^{\mathrm{P}}_{2\mathrm{v}}$ symmetry and are nondegenerate corresponding to different one-dimensional irreps, i.e., $\hat{\mathbf{x}}$ being the basis function of the irrep $\mathbf{1}^{\mathrm{P}}_{2\mathrm{v}}$ and $\hat{\mathbf{y}}$ corresponding to $\tilde{1}^{\mathrm{P}}_{2\mathrm{v}}$. (see Table A.7). Thus, as shown on the right of Fig. 5.2, the associated vector modes are governed by $\mathbf{C}^{\mathrm{S}}_{\infty\mathrm{v}}\otimes\mathbf{C}^{\mathrm{P}}_{2\mathrm{v}}$ symmetry with $e_x\hat{\mathbf{x}}$ and $e_y\hat{\mathbf{y}}$ corresponding to, respectively, LP^x_{lm} and LP^y_{lm} being basis functions of the two-dimensional direct product irreps $\mathbf{1}^{\mathrm{S}}_{\infty\mathrm{v}}\otimes\mathbf{1}^{\mathrm{P}}_{2\mathrm{v}}$ and $\mathbf{1}^{\mathrm{S}}_{\infty\mathrm{v}}\otimes\tilde{\mathbf{1}}^{\mathrm{P}}_{2\mathrm{v}}$.

Note that in terms of the symmetry groups, the weak guidance limit for weak birefringence is similar to the case of introducing ellipticity except that the scalar magnitude and polarization symmetries are reversed from $\mathbf{C}^{\mathrm{S}}_{2\mathrm{v}}\otimes\mathbf{C}^{\mathrm{P}}_{\infty\mathrm{v}}$ for ellipticity.

§ Weak Guidance Limit for Strong Birefringence

From the symmetry viewpoint, the equivalent elliptical guides seen by e_x and e_y in the case of strong birefringence correspond to the fact that Eqs. (5.3*a*) and (5.3*b*) *separately* correspond to $\mathbf{C}^{\mathrm{S}}_{2\mathrm{v}}$ symmetry. Thus, in contrast to the case of weak birefringence, even and odd pairs of scalar solutions e_{xe} and e_{xo} (or e_{ye} and e_{yo}) etc. are nondegenerate; i.e., their orientations are restricted.

Furthermore the total system described by the uncoupled *pair* of Eqs. (5.3) corresponds to $\mathbf{C}^{\mathrm{S}}_{2\mathrm{v}}\otimes\mathbf{C}^{\mathrm{P}}_{2\mathrm{v}}$ symmetry. That Eqs. (5.3) are uncoupled means that the corresponding vector modes are exactly linearly polarized and are given by $e_{xe}\hat{\mathbf{x}}$, etc. Obviously the nondegeneracy of both the scalar magnitude pairs and the polarization vectors means that *even in the weak guidance limit, if strong birefringence is considered, then all four LP modes are nondegenerate.* This weak-to-strong birefringence LP mode splitting corresponds to the symmetry reduction $\mathbf{C}^{\mathrm{S}}_{2\mathrm{v}}\otimes\mathbf{C}^{\mathrm{P}}_{\infty\mathrm{v}} \supset \mathbf{C}^{\mathrm{S}}_{2\mathrm{v}}\otimes\mathbf{C}^{\mathrm{P}}_{2\mathrm{v}}$ which is trivially given by considering the polarization reduction $\mathbf{C}^{\mathrm{P}}_{\infty\mathrm{v}} \supset \mathbf{C}^{\mathrm{P}}_{2\mathrm{v}}$ separately giving $\mathbf{C}^{\mathrm{S}}_{2\mathrm{v}}\otimes(\mathbf{C}^{\mathrm{P}}_{\infty\mathrm{v}} \supset \mathbf{C}^{\mathrm{P}}_{2\mathrm{v}})$. This weak-to-strong birefringence symmetry reduction is shown in Fig. 5.2 as an intermediate or alternative route to the splitting obtained via consideration of Δ in the next subsection. §

5.1.3 Field Component Coupling

Taking account of finite Δ, it is well known that terms $\partial_j \ln n_j^2$ in P_{ij} lead to a correction to the normalized U of order Δ and true a correction to n_{eff} of order $n\Delta^2$. From the point of view of symmetry, the coupling between

e_x and e_y introduced by P_{xy} means that the total system symmetry, i.e., scalar magnitude and polarization, *jointly* must satisfy $\mathbf{C}_{2v}$ symmetry; then the symmetry reduction associated with the coupling of e_x and e_y, that is, $\mathbf{C}^{S}_{\infty v} \otimes \mathbf{C}^{P}_{2v} \supset \mathbf{C}^{J}_{2v}$, leads to the splitting of modes corresponding to even and odd magnitude distributions for each LP polarization, as shown in Fig. (5.2). The fact that Eqs. (5.1) are coupled means that the modes are only exactly LP in the small Δ limit. As Δ is increased, there is an increasing mixing of the polarization components. As in the case of ellipticity, the transition to the TE, TM, HE, and EH modes shown to the left of Fig. 5.2 is determined using the principle of anticrossing together with knowledge of (1) the order of the LP modes which depends on the sign of the birefringence δ_{xy} and (2) the order of the "true" modes which depends on V.

§ In general, an anisotropic fiber can be understood as analogous to two dissimilar parallel isotropic fibers. Just as coupling between the fibers is in general negligible and the supermodes have their power concentrated in just one of the fibers, coupling between the polarization components is usually very small, hence the approximate LP nature of the modes. §

§ Degeneracy between Modes of *x*- and *y*-Polarized Profiles in Isolation

This provides an exception to the approximate LP nature of (linearly) anisotropic fiber modes. Analogous to the case of phase-matched parallel fiber couplers, the modes of the total structure will be *symmetric* and *antisymmetric* combinations of the modes of each profile having the appropriate symmetry; e.g., given degeneracy between LP_{01} of the *y*-polarized profile and LP_{21} of the *x*-polarized profile, modes of the total anisotropic structure are given by the combinations $LP^{y}_{01} \pm LP^{ox}_{21}$ as discussed in Ref. 103. §

5.1.4 Splitting by δ_{xy} of Isotropic Fiber Vector Modes Dominated by Δ-Splitting

As for an elliptical fiber, if we start with the circular isotropic fiber vector modes, splitting of the remaining degeneracies, i.e., for the HE (and EH) pairs, results from consideration of the symmetry reduction $\mathbf{C}^{J}_{\infty v} \supset \mathbf{C}^{J}_{2v}$ shown in Fig. 5.2 as the second reduction from the left.

§ However, in Ref. 104, as opposed to the case of strong birefringence of Sec. 5.1.1, where we showed an "exact" analogy for the *decoupled* field components considered separately, if we rewrite the isotropic fiber VWE with ellipticity as a perturbation [50], then we see that the details of the coupling between e_x and e_y are quite different from the case of anisotropy in Eqs. (5.1). This is not surprising when we consider (1) that *for the strong birefringence decoupled case, each polarization component sees a different equivalent elliptical fiber*, and (2) in light of the above-mentioned scalar-polarization symmetry inversion and thus the different symmetry paths to approach $\mathbf{C}_{2v}^{J}$ from the right of the transition diagrams, that is, $\mathbf{C}_{2v}^{S} \otimes \mathbf{C}_{\infty v}^{P} \supset \mathbf{C}_{2v}^{J}$ for ellipticity, $\mathbf{C}_{\infty v}^{S} \otimes \mathbf{C}_{2v}^{P} (\supset \mathbf{C}_{2v}^{S} \otimes \mathbf{C}_{2v}^{P}) \supset \mathbf{C}_{2v}^{J}$ for anisotropy. §

5.1.5 Correspondence between Isotropic "True" Modes and Birefringent LP Modes

Given the results of Secs. 5.1.3 and 5.1.4, we simply use the principle of *anticrossing* for modes of the same symmetry and then match the irreps to determine the appropriate transformation. Note that this depends on the order of the modes in each splitting which depends on the parameters chosen. In Fig. 5.2, the order of the isotropic fiber modes corresponds to $V < 3.8$, and that of the LP modes to $n_x > n_z > n_y$ for the reasons explained in Sec. 5.1.1.

5.2 RADIAL BIREFRINGENCE

Consider a birefringent fiber with principal axes depending on position but aligned with the radial and azimuthal directions $\hat{\mathbf{r}} = \hat{\mathbf{r}}(\phi)$ and $\hat{\boldsymbol{\phi}} = \hat{\boldsymbol{\phi}}(\phi)$, as introduced in Refs. 44 and 45.

5.2.1 Wave Equations: Longitudinal Invariance

The full wave equation for a longitudinally invariant linearly birefringent fiber is given by the two coupled equations

$$\left\{\nabla_t^2 + k^2 n_r^2 - \beta^2 - \frac{1}{r^2}\right\} e_r = P_{rr} e_r + (P_{r\phi} - C_\phi)\, e_\phi \tag{5.4a}$$

$$\left\{\nabla_t^2 + k^2 n_\phi^2 - \beta^2 - \frac{1}{r^2}\right\} e_\phi = (P_{\phi\phi} + C_\phi) e_r + P_{\phi\phi} e_\phi \tag{5.4b}$$

where

$$C_\phi\, e_j = \frac{2\partial}{r^2 \partial\phi} e_j \tag{5.4c}$$

and the perturbation terms are

$$P_{ij}e_j = \partial_i\left\{2\delta_{zj}\,\frac{1}{r}\,\partial_j(re_j) - (1-2\delta_{zj})\,e_j\,\partial_j(\ln n_j^2)\right\} \qquad \text{and}$$

$$ij = rr,\ r\phi,\ \phi r,\ \text{or}\ \phi\phi \tag{5.4d}$$

$$\partial_r \equiv \frac{\partial}{\partial r} \qquad \partial_\phi \equiv \frac{1}{r}\frac{\partial}{\partial\phi} \qquad 2\delta_{ij} = 1 - \frac{n_j^2}{n_i^2},\ \text{etc.} \tag{5.4e}$$

In general, the terms C_ϕ provide a strong coupling between the radial and azimuthal components of the field given any azimuthal variation of these components, i.e., for any mode of an isotropic fiber except for $\nu = 0$ modes TM_{0m} and TE_{0m}. However, with appropriate choice of n_r sufficiently different from n_ϕ and if the field can be squeezed out of the central region by a core depression or metal core, for example, then reduction of the coupling may be possible and higher-order modes of limiting form $TM_{\nu m}$ and $TE_{\nu m}$ may be supported. This is the case for the ring-type structures in Fig. 5.3,

FIGURE 5.3

Idealized examples of (a, *b*) radially ($\delta_{r\phi} > 0$) and (c, *d*) azimuthally ($\delta_{r\phi} < 0$) anisotropic structures where $2\delta_{r\phi} = 1 - n_\phi^2/n_r^2$. (a, *b*) *Radially anisotropic* ring (a) and cladding (*b*), e.g., liquid crystal molecules aligned perpendicular to the ring interface [44]. (c, *d*) *Azimuthally anisotropic* ring (c) and cladding ring (*d*), e.g., azimuthal anisotropy can be provided by finely spaced concentric ring layers. Guidance for a given polarization is diminished as the ring/cladding interface difference seen by that polarization diminishes (e.g., in part *b* the radial field component is not guided and thus only TE_{0m} modes are guided). Furthermore note thus that (a) and (*d*) act as *radial polarizers*, while (*b*) and (c) act as *azimuthal polarizers*.

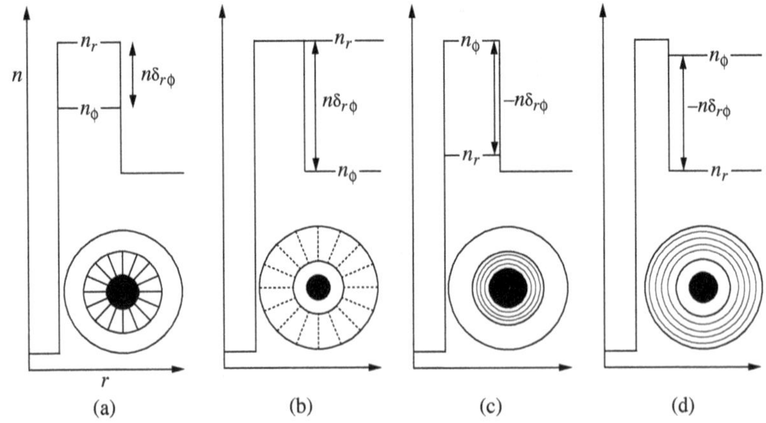

where the field will be essentially zero in the central region and periodic in the ring with electric field polarization perpendicular (TM) or parallel (TE) to the ring interface. While modes of this labeling are supported by a homogeneous dielectric filled coaxial metal guide [105], note that the polarization details of metal and dielectric interfaces differ and that the ring anisotropy plays a crucial role in polarizing the field.

5.2.2 Mode Transitions for Circular Symmetry

If the radial, azimuthal (and longitudinal) field polarization components see indices that retain circular symmetry, that is, $n_i = n_i(r)$, $i = r, \phi$ (and z), then the system is governed by $\mathbf{C}_{\infty v}$ symmetry. Thus, while an effect of the birefringence $\delta_{r\phi}$ is to split TE and TM modes, this is also a consequence of finite Δ. Furthermore, the degeneracies are the same as for the true modes of an isotropic circularly symmetric fiber, as shown in Fig. 5.4. While the birefringence may be such that they are no longer bound modes, the even and odd polarization states of each HE (and EH) mode pair remain degenerate.

5.3 CIRCULAR BIREFRINGENCE

We consider a magneto-optic *gyrotropic medium* [106, 107] with cartesian basis dielectric tensor

$$\mathbb{n}^2\mathbf{E} = n^2 \begin{pmatrix} 1 & -i\delta_{\mathrm{LR}} & 0 \\ -i\delta_{\mathrm{LR}} & 1 & 0 \\ 0 & 0 & \left(\dfrac{n_z}{n}\right)^2 \end{pmatrix} \begin{pmatrix} E_x \\ E_y \\ E_z \end{pmatrix} \tag{5.5a}$$

where $\delta_{\mathrm{LR}} = -2\gamma\,\mu_0 H_z / nk$ with γ being the Verdet coefficient in radians per tesla per meter.

§ We remark that for a medium with *natural optical activity* [76 (p. 79), 83, 84], which strictly speaking appears via the magnetic permittivity tensor, we can use an effective dielectric tensor [82 (Ch. 6)] of the same form as Eq. (5.5*a*) except that $\delta_{\mathrm{LR}} = \chi_m(\beta/|\beta|)$ (1) has the sign depending on the direction of propagation (incorporated with a factor $\beta/|\beta|$) and (2) is a coefficient of the medium via $\chi_m = c\mu_0 M/nE$ (which is the scalar version of the magnetic susceptibility relating the magnetization M to E) and is proportional to the rotary power coefficient of the medium $\Phi = \mathrm{Im}\,\{nk\chi_m/2\}$. §

FIGURE 5.4

Example of level splitting due to azimuthal birefringence $\delta_{r\phi} < 0$ such as in Fig. 5.3c. Note that this schematic is an extreme idealized example and that the ordering of the modes is highly dependent on the form, magnitude, and placement of the anisotropy. Adding $\delta_{r\phi}$ birefringence to Δ-splitting does not change the symmetry from $\mathbf{C}_{\infty\mathbf{v}}$ and thus does not result in splitting of any even/odd modes. However, whether the limiting mode is TE or TM will depend on the sign and placement of the birefringence; e.g., for a radially birefringent ring, the limiting form of HE_{11} will be TM_{11} and the limiting form of EH_{11} will be TE_{11}.

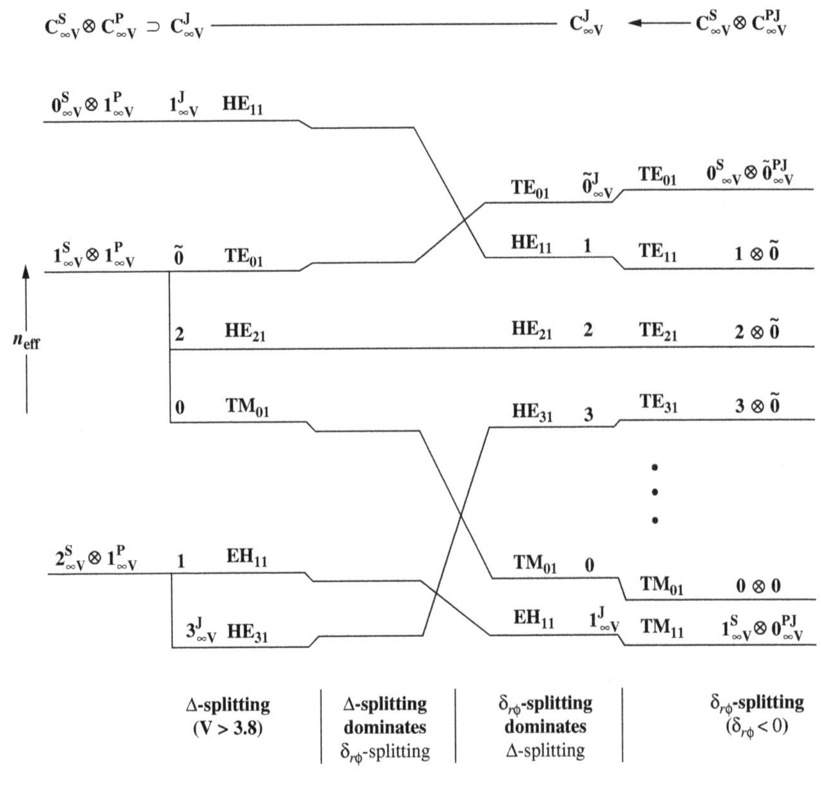

In a circularly polarized basis $\mathbb{n}^2\mathbf{E}$ diagonalizes so that we can associate refractive indices $n_\pm = n(1 \pm \delta_{LR})^{1/2}$ with fields of positive/negative helicity $E_\pm$, that is, in terms of the parameters of Table 2.1 with interpretation discussed in the caption of Fig. 3.2,

$$\mathbb{n}^2\mathbf{E} = n_+^2 E_+ \hat{\mathbf{R}} + n_-^2 E_- \hat{\mathbf{L}} + n_z^2 E_z \hat{\mathbf{z}}$$

$$E_\pm(\mathbf{r}, z) \equiv \frac{1}{\sqrt{2}}(E_x \pm iE_y) = e_\pm(\mathbf{r})e^{i\beta z} \tag{5.5b}$$

5.3.1 Wave Equation

For circularly birefringent refractive indices as in Eq. (5.5) from the general result of Eq. (2.11) we obtain the wave equation

$$\{\nabla_t^2 + k^2 n_+^2 - \beta^2\} e_+ = P_{++} e_+ + P_{+-} e_- \tag{5.6a}$$

$$\{\nabla_t^2 + k^2 n_-^2 - \beta^2\} e_- = P_{-+} e_+ + P_{--} e_- \tag{5.6b}$$

with

$$P_{++} e_+ = \partial_+ \{2\delta_{z+} \partial_- e_+ - (1 - 2\delta_{z+})\, e_+ \partial_- (\ln n_+^2)\} \tag{5.7a}$$

$$P_{+-} e_- = \partial_+ \{2\delta_{z-}\, \partial_+ e_- - (1 - 2\delta_{z-})\, e_- \partial_+ (\ln n_+^2)\} \tag{5.7b}$$

$$P_{-+} e_+ = \partial_- \{2\delta_{z+}\, \partial_- e_+ - (1 - 2\delta_{z+})\, e_+ \partial_- (\ln n_+^2)\} \tag{5.7c}$$

$$P_{--} e_- = \partial_- \{2\delta_{z-}\, \partial_+ e_- - (1 - 2\delta_{z-})\, e_- \partial_+ (\ln n_+^2)\} \tag{5.7d}$$

where

$$\partial_\pm \equiv \frac{1}{\sqrt{2}} \left\{ \frac{\partial}{\partial x} \pm i \frac{\partial}{\partial y} \right\} = \frac{1}{\sqrt{2}} e^{i\phi} \left\{ \frac{\partial}{\partial \rho} \pm \frac{i}{\rho} \frac{\partial}{\partial \phi} \right\} \quad \text{and}$$

$$\delta_{z\pm} = \frac{1}{2} \left(1 - \frac{n_\pm^2}{n_z^2} \right) \tag{5.8}$$

In the weak guidance limit for which the RHS circular polarization coupling terms vanish, Eqs. 5.6*a* and *b* are decoupled and are equivalent to the standard coupled linearly polarized mode formalism as in Ref. 89.

5.3.2 Symmetry and Mode Splittings

When circular birefringence $\delta_{LR} = (n_L - n_R)/n$ is introduced, σ_v is no longer a symmetry operation of the system (i.e., the polarization reflection $O_p(\sigma_v)\, \hat{\mathbf{L}} = \hat{\mathbf{R}}$ would mean conversion between now nondegenerate polarizations). Thus the symmetry is reduced from that of the rotation-reflection group $\mathbf{C}_{\infty v}$ to that of pure rotation group $\mathbf{C}_\infty$ (sometimes referred to as $\mathbf{SO}_2$—see Table A.1); the consequence of the reduction $\mathbf{C}_{\infty v} \supset \mathbf{C}_\infty$ (or equivalently $\mathbf{O}_2 \supset \mathbf{SO}_2$) that occurs for polarization and thus joint symmetries is a splitting of the doubly degenerate isotropic fiber vector modes HE and EH as shown in Fig. 5.5. On the right of that figure, analogous to the linearly

FIGURE 5.5

Level splitting due to circular birefringence $2\delta_{LR} = 1 - n_R^2/n_L^2$.

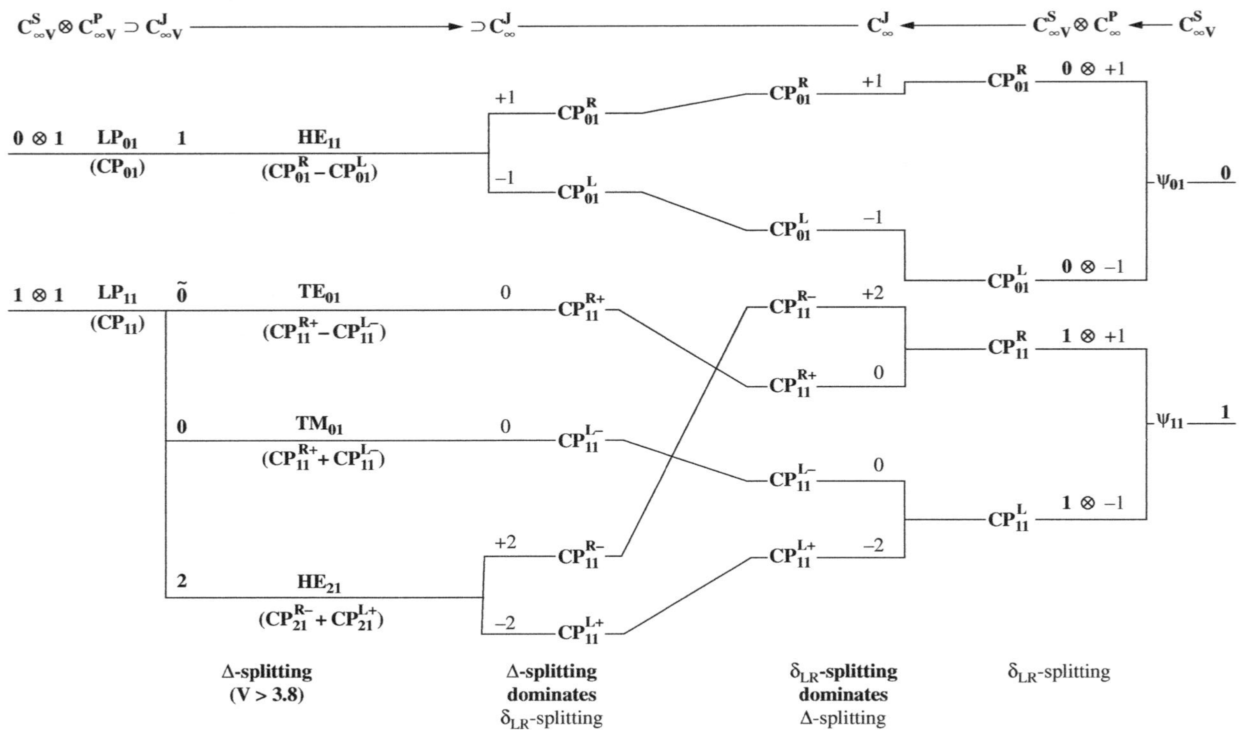

birefringent case, we also see a splitting between the left (L) and right (R) circularly polarized modes (CP) modes corresponding to the appropriate VWE to zeroth order in Δ for weak birefringence. Moving one column to the left in that figure, we then see a finer splitting between the "+" and "−" modes of the L and R pairs when we consider the effect of finite Δ, i.e., the reduction to joint symmetry.

§ In fact, analogous to the case of strong linear birefringence mentioned in Sec. 5.1.1, the latter +/− splitting would also be seen even while ignoring Δ effects if we considered strong circular birefringence. §

CHAPTER 6

Multicore Fibers and Multifiber Couplers

In this chapter, we consider the scalar and vector supermodes of structures consisting of weakly coupled single- or few-mode fiber cores (or fibers) forming arrays with discrete rotational symmetry.

Details of these structures are given in Sec. 6.1. In Sec. 6.2 we summarize general results for degeneracies and give basis functions that can be used directly for construction of both scalar modes (Sec. 6.3) and vector modes (Sec. 6.4) to obtain the general form of n-core fiber fields. Modal propagation constant degeneracies and example field constructions are also given. Section 6.5 discusses numerical evaluation, and Sec. 6.6 treats propagation constant splittings. Apart from their possible polarizing nature, from the practical viewpoint the major interest in multicore fibers and multifiber couplers is due to the power transfer that occurs between the constituent guides. This may be analyzed in terms of supermode interference as discussed in Sec. 6.7.

6.1 MULTILIGHTGUIDE STRUCTURES WITH DISCRETE ROTATIONAL SYMMETRY

Here our objective is to give general principles applicable to a wide range of multiguide structures possessing discrete rotational symmetry.

6.1.1 Global $\mathbf{C}_{nv}$ Rotation-Reflection Symmetric Structures: Isotropic Materials

In our application examples, we concentrate on structures with multiple circular isotropic cores (or fibers) that are placed equidistant so as to form a "ring" array with discrete n-fold rotation-reflection symmetry $\mathbf{C}_{nv}$. For example, the case of three circular cores is shown in Fig. 6.1*a* which also provides the coordinate system for the study.

In general, rotation reflection symmetric multiguide structures can include component guides with individual symmetry reduced to $\mathbf{C}_{iv}$ ($i = 1, \ldots, \infty$) as long as each guide has a symmetry plane passing through the fiber center and corresponding to the n reflection symmetry planes $(x_i - z)$ required for global symmetry $\mathbf{C}_{nv}$. Figure 6.1*b* considers the case of global $\mathbf{C}_{3v}$ symmetry with individual core symmetry $\mathbf{G}_i = \mathbf{C}_{1v}$.

The $\mathbf{C}_{nv}$ symmetric structures can also include a central core and multiple rings of cores, provided that each has at least an integer multiple of n-fold $\mathbf{C}_{nv}$ symmetry with the symmetry planes coinciding. The $\mathbf{C}_{nv}$ symmetry always allows us to place constraints on the supermode degeneracies and the global azimuthal dependence of their fields and thus the relation between the field in each of the

FIGURE 6.1

Crosssection of the isotropic three-core fiber with cores arranged to have discrete global three-fold rotation-reflection symmetry $\mathbf{C}_{3v}$ about the fiber center, i.e., equilateral triangular placement.

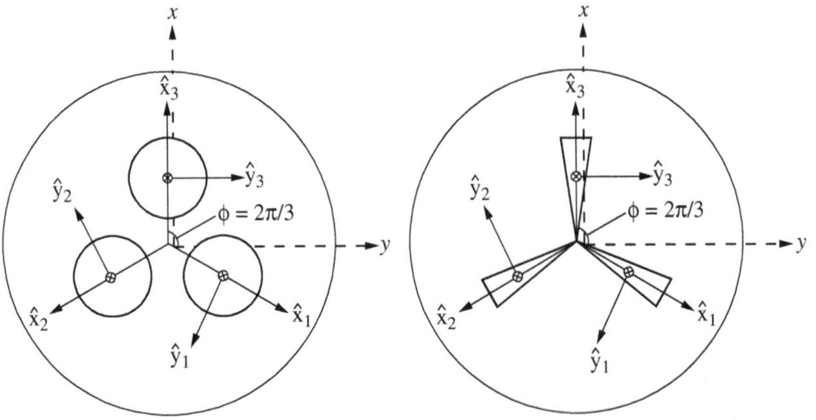

(*a*) $\mathbf{G}_g = \mathbf{C}_{3v}, \mathbf{G}_i = \mathbf{C}_{\infty v}$ (*b*) $\mathbf{G}_g = \mathbf{C}_{3v}, \mathbf{G}_i = \mathbf{C}_{1v}$

n-fold sectors of the guide. However, as the radial "symmetry" diminishes from that of a single ring of pointlike cores, the accuracy of approximate analyses in terms of individual core fields rapidly diminishes, and thus one simply uses the azimuthal symmetry to simplify a general numerical analysis. In particular, $\mathbf{C}_{nv}$ places no constraints on the radial field dependence. Thus, for example, apart from the requirement that only central and ring modes of similar azimuthal dependence can combine as supermodes, it places no constraints on the relative amplitudes of the field in a central core and successive rings.

6.1.2 Global C_{nv} Symmetry: Material and Form Birefringence

The analysis can also be adapted to cores composed of anisotropic materials; *Global linear anisotropy* corresponds to the birefringent axes of each core arranged in a common direction with respect to the global cartesian axes. *Discrete global radial anisotropy* or $\mathbf{C}_{nv}$ anisotropic structures are obtained for n cores arranged with geometric $\mathbf{C}_{nv}$ symmetry and each having the same birefringence but with a principal axis aligned toward the global fiber center.

Examples of such global linear and radial birefringence are provided by (1) 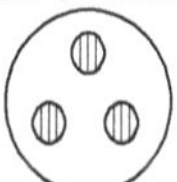 and (2) 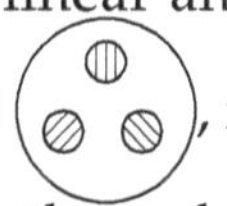, respectively. Both have $\mathbf{C}_{3v}$ global scalar symmetry seen by the scalar modes. However, the linear case 1 has polarization symmetry $\mathbf{C}_{2v}$, whereas the radial case 2 has polarization symmetry $\mathbf{C}_{nv}$.

Forms of *discrete global radial anisotropy* can also be realized *geometrically* as in the examples of Fig. 6.2. We note that some of these structures with high azimuthal nonuniformity (particularly the cases $n = 4$ which have an "iron cross" form as well as $n = 8$ analogs corresponding to octopus-type structures) have been of particular interest as candidates for obtaining circularly birefringent fibers upon twisting [108]. See Refs. 109 to 111 for design principles and analytical solutions for the scalar fields.

6.1.3 Global C_n Symmetric Structures

If identical cores have cross sections with no symmetry axes or their symmetry axes are tilted by the same angle with respect to axes to the fiber center, then although the total structure may remain

FIGURE 6.2

Examples of structures with discrete global radial anisotropy and $\mathbf{C_{nv}}$ symmetry. The dark central regions represent central depressions that act to squeeze the field out of the fiber center and thus aid the decoupling of its radial and azimuthal components.

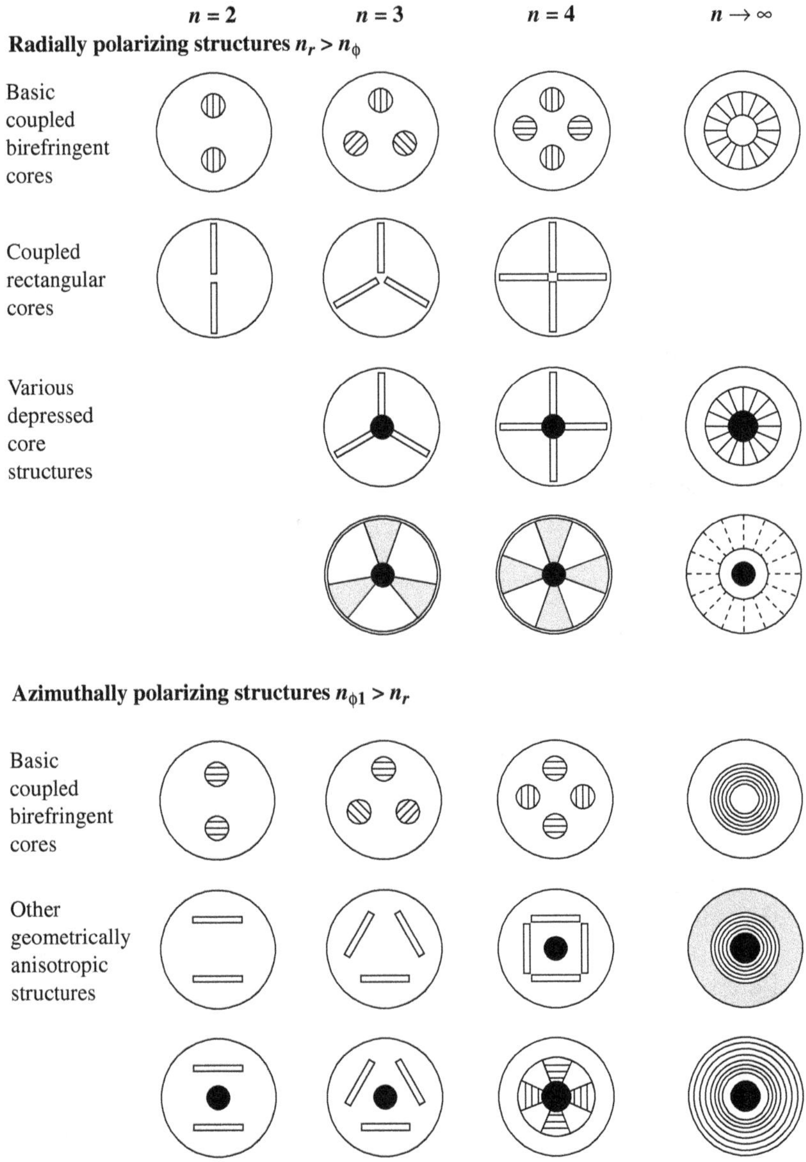

invariant under n-fold rotations, it is no longer invariant under reflection. Thus the global symmetry is reduced to $\mathbf{C}_n$, for example,  corresponds to global scalar symmetry $\mathbf{C}_4$ with individual core scalar symmetry $\mathbf{C}_{2v}$. Four identical anisotropic circular cores with each core having corresponding principal axes tilted by the same angle with respect to the axes to the fiber center would correspond to global scalar symmetry $\mathbf{C}_{4v}$ and global polarization symmetry $\mathbf{C}_4$ (with individual core scalar symmetry $\mathbf{C}_{\infty v}$ and individual polarization symmetry $\mathbf{C}_{2v}$).

6.2 GENERAL SUPERMODE SYMMETRY ANALYSIS

The consequence of symmetry for the multiguide structure is that the normal modes of the total structure, which are normally referred to as "supermodes," may be constructed from the field solution in certain sectors of the total structure with appropriate boundary conditions, thus reducing the computation required in a general numerical calculation. Furthermore, a good approximation is usually provided by construction of the supermodes in terms of the modes of the individual guides.

Before outlining the general procedure for supermode construction (in Sec. 6.6.2), we first consider (in Sec. 6.2.1) qualitative determination of the supermode degeneracy due to the imposed symmetry. Those readers whose major interest is in the resulting supermode forms may prefer to proceed directly to Sec. 6.3 and refer to these sections at a later stage.

6.2.1 Propagation Constant Degeneracies

General Analysis for Global n-Core Fibers with Global C_{nv} Symmetry

Given an individual core j with symmetry $\mathbf{C}_{iv}$ ($i = 1, \ldots, \infty$), consider the modes $\bar{\psi}_j^s$ of that core in isolation transforming as irrep $\mathbf{l}(\mathbf{C}_{iv})$ and having degeneracy $|\mathbf{l}|$. When n such cores are placed in an array of symmetry $\mathbf{C}_{nv}$, ignoring the interaction, we get a set of $n|\mathbf{l}|$ degenerate isolated core modes

$$\{\{\bar{\psi}_j^s: s = 1, \ldots, |l|\}_j, j = 1, \ldots, n\} \qquad (6.1)$$

It is possible to determine the degeneracy breaking for the normal modes of the array, i.e., the supermodes, by symmetry arguments only; e.g., see Ref. 5 for an excellent tutorial discussion of the procedure.

As the action of symmetry operation $O(g)$ corresponding to an element $g \in \mathbf{C_{nv}}$ is simply to transfer the field from one core to another, [for example, $O(g)\,\bar{\psi}_i^s = \bar{\psi}_j^s$ transferring mode s from core i to j], one can construct a set of matrices $\mathbb{M}^{(\mathbf{n},l)}(g)$ that perform this transfer function on all the modes:

$$O(g)\left[\bar{\psi}_1^1,\ldots,\bar{\psi}_j^s,\ldots,\bar{\psi}_n^{|l|}\right]=\left[\bar{\psi}_1^1,\ldots,\bar{\psi}_j^s,\ldots,\bar{\psi}_n^{|l|}\right]\mathbb{M}^{(\mathbf{n},l)}(g) \qquad (6.2)$$

However, it is simpler to work with matrices that act on symmetry adapted linear combinations (SALCs) of the individual core modes or normal modes, i.e., matrices for which the supermodes are basis functions. These matrices are *block diagonal* (cf. Sec. A.3.1) and given by a transformation of the form $\mathbb{T}\,\mathbb{M}^{(\mathbf{n},l)}(g)\,\mathbb{T}^{-1}$. For example,

$$O(g)\left[\bar{\Psi}_1^{(0)},\ldots,\bar{\Psi}_S^{(L)},\ldots\right]=\left[\bar{\Psi}_1^{(0)},\ldots,\bar{\Psi}_S^{(L)},\ldots\right]\begin{bmatrix}\mathbb{D}^{(0)}(g) & & & & \\ & \mathbb{D}^{(1)}(g) & & & \\ & & \boxed{\cdot} & & \\ & & & \boxed{\cdot} & \\ & & & & \boxed{\cdot}\end{bmatrix} \qquad (6.3)$$

Thus, the set of matrices $\mathbb{M}^{(\mathbf{n},l)}(g)$ corresponds to a matrix representation of $\mathbf{C_{nv}}$ which we denote as $\mathbf{nl}\,(\mathbf{C_{nv}})$ and which, being of dimension $n\,|\mathbf{l}|$, is necessarily reducible to the irreducible representations of $\mathbf{C_{nv}}$, that is, the irreps $\mathbf{L(C_{nv})}$.

However, it is not necessary to determine $\mathbb{T}$ or even the details $\mathbb{M}^{(\mathbf{n},l)}(g)$ to obtain the reduction. Following a standard procedure, particularly well explained in Ref. 5, for example, one simply obtains the diagonal elements $\mathbb{M}^{(\mathbf{n},\mathbf{l})}(g)_{jj}$ and thus the characters $\chi^{(\mathbf{n},\mathbf{l})}(g)$ which are then substituted into the appropriate form of Eq. (A.6). This leads to the result [2] that the decomposition of the reducible representations $\mathbf{nl}\,(\mathbf{C_{nv}})$ to the irreps $\mathbf{L(C_{nv})}$ is of the form

$$\boxed{\begin{aligned}\mathbf{nl}(\mathbf{C_{nv}}) \rightarrow [\mathbf{0} \oplus \mathbf{1} \oplus \cdots \oplus \mathbf{L_n} \oplus (\mathbf{n}/\mathbf{2})_{\mathrm{n=even}}]_{\mathrm{l}\neq\tilde{0},\,\widetilde{\mathrm{i}/2}} \\ \oplus[\tilde{\mathbf{0}} \oplus \mathbf{1} \oplus \cdots \oplus \mathbf{L_n} \oplus \widetilde{(\mathbf{n}/\mathbf{2})}_{\mathrm{n=even}}]_{\mathrm{l}\neq 0,\,\mathrm{i}/2}\end{aligned}} \qquad (6.4)$$

with (1) the terms in brackets appearing if the conditions in the subscript are satisfied and (2) L_n being the largest integer smaller than $n/2$, that is, $L_n = (n/2) - 1$ for n even and $(n-1)/2$ for n odd.

In general, the irreps $\mathbf{0}$, $\tilde{\mathbf{0}}$, $(\mathbf{n/2})$, and $\widetilde{(\mathbf{n/2})}$, being of dimension 1, correspond to nondegenerate supermodes; the other irreps $(\mathbf{1}, \ldots, \mathbf{L}_n)$ being of dimension 2 correspond to doubly degenerate supermodes.

Furthermore, note that all irreps $\mathbf{L(C_{nv})}$ appear at least once in the reduction of the reducible representations $\mathbf{nl(C_{nv})}$ except for the cases of the isolated core modes transforming as $\mathbf{l(C_{iv})} = \mathbf{0}, \tilde{\mathbf{0}}, (\mathbf{i/2})$, or $\widetilde{(\mathbf{i/2})}$.

Note also that the number of supermodes generated simply corresponds to the total number of isolated core modes, i.e., the dimension $n|\mathbf{l}|$ of the reducible representations. Thus, e.g., in the case of n circular cores, $2n$ scalar supermodes (n for $\mathbf{l} = 0$) can be supported by the structure for a given isolated core azimuthal mode number l (and radial mode number m).

Example Application: Scalar Supermode Degeneracies for Three Circular Cores with Global C_{3v} Symmetry ($n = 3$, $i = \infty$)

For example, for $n = 3$, $i = \infty$, the set of degenerate modes of individual isolated cores is

$$\{F_l(r_j): j = 1, \cdots, 3\} \qquad l = 0 \tag{6.5a}$$

$$\{F_l(r_j)\cos l\phi_j, F_l(r_j)\sin l\phi_j: j = 1, \cdots, 3\} \qquad l \rhd 0 \tag{6.5b}$$

The degeneracy splitting is then given by noting that in Eq. (6.4) the terms $(\mathbf{n/2})$ and $\widetilde{(\mathbf{n/2})}$, being nonintegral, do not appear, and thus for $l = 0$

$$\mathbf{30} \rightarrow \mathbf{0} \oplus \mathbf{1} \tag{6.6a}$$

For $\boldsymbol{l} > \mathbf{0}$, we include the bracketed sequence up to $\mathbf{L_n} = \mathbf{1}$, giving

$$\mathbf{31} \rightarrow \mathbf{0} \oplus \tilde{\mathbf{0}} \oplus \mathbf{1} \oplus \mathbf{1} \qquad (l > 0) \tag{6.6b}$$

Physically, Eq. (6.6*b*) tells us that the six degenerate modes in Eq. (6.5*b*) are split into four sets of supermodes with each supermode in a set having the same propagation constant. Among these sets, two are nondegenerate and two are doubly degenerate [because $\mathbf{1(C_{3v})}$ is of dimension 2]. A similar interpretation stands for Eq. (6.6*a*). The unique vector basis of $\mathbf{0}$ is even whereas the one for $\tilde{\mathbf{0}}$ is odd.

§ Degeneracies for n-Core Fibers with Global C_n Symmetry

In the case of global $\mathbf{C_n}$ Symmetry (see Sec. 6.1.3) and isolated core modes again transforming as $l(\mathbf{C_{iv}})$ the reduction of the *reducible* representations **nl** of $\mathbf{C_n}$ [corresponding to the transfer matrices for the isolated core modes analogous to those in Eqs. (6.2) and (6.3)] to the *irreps* of $\mathbf{C_n}$ takes the form

$$\mathbf{n}l(\mathbf{C_n}) \rightarrow [\mathbf{0} \oplus \pm\mathbf{1} \oplus \cdots \oplus \pm\mathbf{L_n} \oplus (\mathbf{n/2})_{\mathrm{n=even}}] \\ \oplus [\mathbf{0} \oplus \pm\mathbf{1} \oplus \cdots \oplus \pm\mathbf{L_n} \oplus (\mathbf{n/2})_{\mathrm{n=even}}]_{\mathrm{l \neq 0,\, i/2}} \tag{6.7}$$

This can be seen from the analogous $\mathbf{C_{nv}}$ result of Eq. (6.4) by simply noting the branching rules for $\mathbf{C_{nv}} \supset \mathbf{C_n}$. In particular, reflection no longer being a symmetry operation means that (1) the irreps $\tilde{\mathbf{0}}$ and $\widetilde{(\mathbf{n/2})}$ are no longer distinguished from **0** and **(n/2)**, respectively, to which they branch, and (2) the two-dimensional irreps $\mathbf{L(C_{nv})}$ branch to $\pm\mathbf{L(C_n)}$, where $\pm\mathbf{L}$ represents the two one-dimensional irreps $+\mathbf{L}$ and $-\mathbf{L}$. These are usually lumped together as an effective two-dimensional irrep as, although no symmetry operation of $\mathbf{C_n}$ will mix the basis functions of these irreps, invariance of the wave equation under inversion of z means that the associated modes remain degenerate. However, in cases of $\mathbf{C_{nv}}$ being lowered to $\mathbf{C_n}$ symmetry by some longitudinal variation which eliminates the σ_v reflection plane and simultaneously destroys the z-reversal symmetry, e.g., a helicoidal twisting, these modes will split. §

6.2.2 Basis Functions for General Field Construction

Given global symmetry group $\mathbf{G} = \mathbf{C_{nv}}$ or $\mathbf{C_n}$, the field of the structure can be decomposed as a linear combination of eigenfunctions associated with the irreps **L(G)**. These eigenfunctions can be generated using a standard and formal procedure based on the projection operator [7 (Sec. 5.1)]

$$P_{ij}^{(L)} = \frac{|L|}{|G|}\sum_g D^{(L)}(g)_{ij}^{*}\, O(g) \tag{6.8}$$

where $|L|$ is the dimension of the irrep **L(G)**, $O(g)$ is the symmetry operator (rotation, reflection, etc.) corresponding to the group element g, and $|G|$ is the group order, i.e., $|\mathbf{C_{nv}}| = 2n$ and $|\mathbf{C_n}| = n$. This projection operator is used to construct functions having the appropriate symmetry as

$$\boxed{\xi_i^{(L)} = \mathrm{P}_{ij}^{(L)}\,\overline{\xi}_j \quad \text{where} \quad \overline{\xi}_j = \mathrm{O}(\mathrm{C}_n^j)\,\overline{\xi}_1} \tag{6.9}$$

and where $\overline{\xi}_j$ is a function attached to the jth coordinate axis. We refer to $\xi_i^{(L)}$ as a *symmetry adapted linear combination* (SALC) or a global field function and to $\overline{\xi}_j$ as the jth core or sector field function.

Symmetry Adapted Linear Combinations for C_{nv} Symmetry

For $\mathbf{C}_{nv}$ global symmetry, we can choose the global field to be even or odd with respect to an n-fold symmetry axis. Furthermore, for application of the reflection symmetry operator it is convenient to decompose the sector field function $\overline{\xi}_j$ into an even and/or an odd part [71, 73], enumerating the following cases [2]

$$\begin{aligned} \overline{\xi}_j &= \overline{\xi}_j^e && (a)\\ \overline{\xi}_j &= \overline{\xi}_j^o && (b)\\ \overline{\xi}_j &= \overline{\xi}_j^e \pm \overline{\xi}_j^o && (c) \end{aligned} \tag{6.10}$$

Substituting into Eq. (6.9) thus generates the set of unnormalized symmetry adapted linear combinations or *basis functions* given in Table 6.1.

§ SALCs for C_n Symmetry

For $\mathbf{C}_n$ global symmetry we have:

$$\xi^{(0)} = \sum_{j=1}^{n} \overline{\xi}_j \qquad L = 0 \tag{6.11a}$$

$$\xi^{(\pm L)} = \sum_{j=1}^{n} \exp\left(\pm i\frac{2\pi Lj}{n}\right)\overline{\xi}_j \quad 1 < \mathrm{L} < \frac{\mathrm{n}}{2} \tag{6.11b}$$

and *if n is even,*

$$\xi^{(n/2)} = \sum_{j=1}^{n} (-1)^j\,\overline{\xi}_j \qquad \mathrm{L} = \frac{\mathrm{n}}{2} \tag{6.11c}$$

The SALCs $\xi^{(\pm L)}$ correspond to *circulating* superfields with the field amplitude in each core around the ring of cores being $\pm 2\pi L/n$ out of phase with that of the preceding core. The extra degeneracy

TABLE 6.1

Symmetry Adapted Linear Combinations for C_{nv} Global Symmetry

(a) $\overline{\xi}_j = \overline{\xi}_j^e$	(b) $\overline{\xi}_j = \overline{\xi}_j^o$	(c) $\overline{\xi}_j = \overline{\xi}_j^e \pm \overline{\xi}_j^o$	
$\xi_e^{(0)} \propto \sum_{j=1}^{n} \xi_j^e$		$\xi_e^{(0)} \propto \sum_{j=1}^{n} \xi_j^e$	$L = 0$
	$\xi_o^{(\bar{0})} \propto \sum_{j=1}^{n} \overline{\xi}_j^o$	$\xi_o^{(\bar{0})} \propto \sum_{j=1}^{n} \overline{\xi}_j^o$	$L = \tilde{0}$
		$\boxed{\overline{\xi}_j = \overline{\xi}_j^e + \overline{\xi}_j^o}$	
$\xi_{e1}^{(L)} \propto \sum_{j=1}^{n} \cos\left(\frac{2\pi Lj}{n}\right)\overline{\xi}_j^e$		$\xi_{e+}^{(L)} = \xi_{e1}^{(L)} + \xi_{e2}^{(L)} \propto \sum_{j=1}^{n} \cos\left(\frac{2\pi Lj}{n}\right)\overline{\xi}_j^e - \sin\left(\frac{2\pi Lj}{n}\right)\overline{\xi}_j^o$	$1 < L < \frac{n}{2}$
$\xi_{o1}^{(L)} \propto \sum_{j=1}^{n} \sin\left(\frac{2\pi Lj}{n}\right)\overline{\xi}_j^e$		$\xi_{o+}^{(L)} = \xi_{o1}^{(L)} + \xi_{o2}^{(L)} \propto \sum_{j=1}^{n} \sin\left(\frac{2\pi Lj}{n}\right)\overline{\xi}_j^e + \cos\left(\frac{2\pi Lj}{n}\right)\overline{\xi}_j^o$	
		$\boxed{\overline{\xi}_j = \overline{\xi}_j^e + \overline{\xi}_j^o}$	
	$\xi_{e2}^{(L)} \propto -\sum_{j=1}^{n} \sin\left(\frac{2\pi Lj}{n}\right)\overline{\xi}_j^o$	$\xi_{e-}^{(L)} = \xi_{e1}^{(L)} - \xi_{e2}^{(L)} \propto \sum_{j=1}^{n} \cos\left(\frac{2\pi Lj}{n}\right)\overline{\xi}_j^e + \sin\left(\frac{2\pi Lj}{n}\right)\overline{\xi}_j^o$	$1 < L < \frac{n}{2}$
	$\xi_{o2}^{(L)} \propto \sum_{j=1}^{n} \cos\left(\frac{2\pi Lj}{n}\right)\overline{\xi}_j^o$	$\xi_{o-}^{(L)} = \xi_{o1}^{(L)} - \xi_{o2}^{(L)} \propto \sum_{j=1}^{n} \sin\left(\frac{2\pi Lj}{n}\right)\overline{\xi}_j^e - \cos\left(\frac{2\pi Lj}{n}\right)\overline{\xi}_j^o$	
And if n Is Even			For n Even
$\xi_e^{(n/2)} \propto \sum_{j=1}^{n} (-1)^j \xi_j^e$		$\xi_e^{(n/2)} \propto \sum_{j=1}^{n} (-1)^j \xi_j^e$	$\boldsymbol{L = \left(\frac{n}{2}\right)}$
	$\xi_o^{(\widetilde{n/2})} \propto \sum_{j=1}^{n} (-1)^j \overline{\xi}_j^o$	$\xi_o^{(\widetilde{n/2})} \propto \sum_{j=1}^{n} (-1)^j \overline{\xi}_j^o$	$\boldsymbol{L = \widetilde{\left(\frac{n}{2}\right)}}$

between $\xi^{(+L)}$ and $\xi^{(-L)}$ related to z-reversal symmetry, as mentioned in the last subsection in Sec. 6.2.1 "Degeneracies for n-Core Fibers with Global $\mathbf{C_n}$ Symmetry" means that the superfields circulating in two opposing directions can be combined to obtain a sinusoidal dependence for the relative amplitudes similar to that for $\mathbf{C_{nv}}$ symmetry in Table 6.1*a* and *b*. However, as reflection is no longer a symmetry operation, the sector field functions ξ_j are no longer decomposed into an even or odd part, and the superfields are neither even nor odd with respect to the n-fold axes. §

SALCs as Approximations to Supermode Fields

In general, given a set of core field functions $\bar{\xi}_j$, the SALCs ξ_i^L generated therefrom provide a set of basis functions for construction of supermode fields as discussed in Sec. 6.4. However, for appropriate choice of core field functions $\bar{\xi}_j$, as modes of the individual cores in isolation, the SALCs ξ_i^L will often individually provide excellent approximations to the supermode fields.

In the following sections we consider the generation of three types of approximate supermodes by considering three cases for the individual core eigenfunctions $\bar{\xi}_j$, that is, individual core scalar modes, true vector modes, and LP vector modes. We also discuss the consequences of symmetry in the simplification of numerical determination of the field.

6.3 SCALAR SUPERMODE FIELDS

First we consider approximate supermode constructions as combinations of individual core modes. Then we discuss more exact treatments and numerical analyses that take into account the dependence of scalar supermode form on core separation. The initial analysis in terms of individual core modes assumes immersion in an infinite cladding. Nevertheless, the supermode form for the $\mathbf{C_{nv}}$ multicore structures may be compared with the mode forms deduced for single-core $\mathbf{C_{nv}}$ structures in Chap. 4.

6.3.1 Combinations of Fundamental Individual Core Modes

For cases of $\mathbf{C_{nv}}$ structure scalar supermodes constructed from individual core fundamental modes, we substitute

$$\boxed{\bar{\xi}_j = \bar{\xi}_j^e = \bar{\psi}_{01}(r_j)} \qquad (6.12)$$

in Table 6.1*a* to obtain the supermodes of Table 6.2.

TABLE 6.2

Fundamental Mode Combination Supermodes

(*a*) L = 0	
1 Fundamental supermode	$\Psi_{01} = \xi^{(0)} = \sum_{j=1}^{n} \bar{\psi}_{01}(r_j)$
(*b*) $1 \le L < \frac{n}{2}$	
L_n Degenerate supermode pairs	(L_n = largest integer smaller than $n/2$)
$\Psi^e_{L1} = \xi^{(L)}_e = \sum_{j=1}^{n} \cos\frac{2\pi Lj}{n}\,\bar{\psi}_{01}(r_j)$	$\Psi^o_{L1} = \xi^{(L)}_o = \sum_{j=1}^{n} \sin\frac{2\pi Lj}{n}\,\bar{\psi}_{01}(r_j)$
(*c*) For n even, $L = \frac{n}{2}$:	
1 Nondegenerate alternating-sign supermode	$\Psi^e_{(n/2),1} = \xi^{(n/2)}_e = \sum_{j=1}^{n} (-1)^j\,\bar{\psi}_{01}(r_j)$

There are n scalar supermodes in total.

As examples, Figs. 6.3 and 6.4 give the scalar supermodes corresponding to individual fiber mode symmetry $lm = 01$ for the case of two, three, four, five, and six cores, respectively.

6.3.2 Combinations of Other Nondegenerate Individual Core Modes

The general form given by Table 6.2 (or 6.1*a*) also applies to supermodes constructed from any nondegenerate individual core mode $\bar{\psi}^e$ that is even with respect to the individual core mirror plane $(x_i\, z)$. For nondegenerate odd modes we use Table 6.1*b*.

6.3.3 Combinations of Degenerate Individual Core Modes

For degenerate individual core modes (e.g., all circular core modes with $l > 0$), we use

$$\bar{\xi}_j = \bar{\xi}^e_j \pm \bar{\xi}^o_j = \bar{\psi}^e_{lm}(r_j, \phi_j) \pm \bar{\psi}^o_{lm}(r_j, \phi_j) \qquad \text{in Table 6.1}c \tag{6.13}$$

which results in $2n$ scalar supermodes. We see Ref. 2 for examples.

FIGURE 6.3

Scalar supermodes for $n = 2, 3, 4$ cores in an array with $\mathbf{C_{nv}}$ symmetry constructed as linear combinations of isolated core modes with $lm = 01$. Field amplitudes in each core and propagation constant corrections assume weakly coupled cores in an infinite cladding.

$n = 2$	$n = 3$	$n = 4$
$\Psi_{01} = \sqrt{2}\,\hat{\Psi}_{01}$ $\delta_1\beta_{01} = C(d)$	$\Psi_{01} = \sqrt{3}\,\hat{\Psi}_{01}$ $\delta\beta_{01} = 2C(d)$	$\Psi_{01} = 2\hat{\Psi}_{01}$ $\delta\beta_{01} = 2C(d) - C(\sqrt{2}d) \approx 2C(d)$
$\Psi_{11}^e = \sqrt{2}\,\hat{\Psi}_{11}^e$ $\delta\beta_{11}^e = -C(d)$	$\Psi_{11}^e = \sqrt{6}\,\hat{\Psi}_{11}^e \qquad \Psi_{11}^o = \sqrt{2}\,\hat{\Psi}_{11}^o$ $\delta\beta_{11}^e = \delta\beta_{11}^o = -C(d)$	$\Psi_{11}^e = \sqrt{2}\,\hat{\Psi}_{11}^e \qquad \Psi_{11}^o = \sqrt{2}\,\hat{\Psi}_{11}^o$ $\delta\beta_{11}^e = \delta\beta_{11}^o = -C(\sqrt{2}d) \approx 0$
		$\Psi_{21}^e = 2\hat{\Psi}_{21}^e$ $\delta\beta_{21}^e = -2C(d) + C(\sqrt{2}d) \approx -2C(d)$

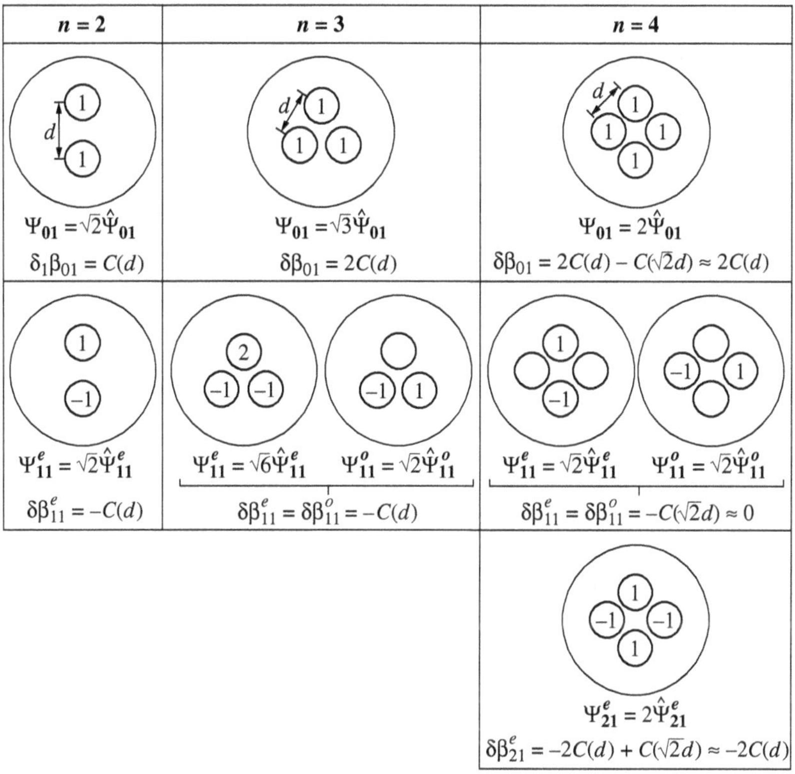

6.4 VECTOR SUPERMODE FIELDS

6.4.1 Two Construction Methods

As illustrated in branches A and B of Fig. 6.5, one may use symmetry in two ways for constructing isotropic fiber vector supermodes: (A) direct application of the projection operator constructions of Sec. 6.2.2 to the individual core vector modes using the substitutions of Table 6.3 and (B) weak-guidance construction in terms of the scalar supermodes as in Table 6.4 (this is analogous to the vector mode constructions in terms of scalar modes in Chaps. 3 and 4). For fundamental mode combinations the two methods lead to the

FIGURE 6.4

Scalar supermodes and their propagation constant corrections $\delta\beta$ for $n = 5$ and $n = 6$ cores constructed from isolated core modes with $lm = 01$.

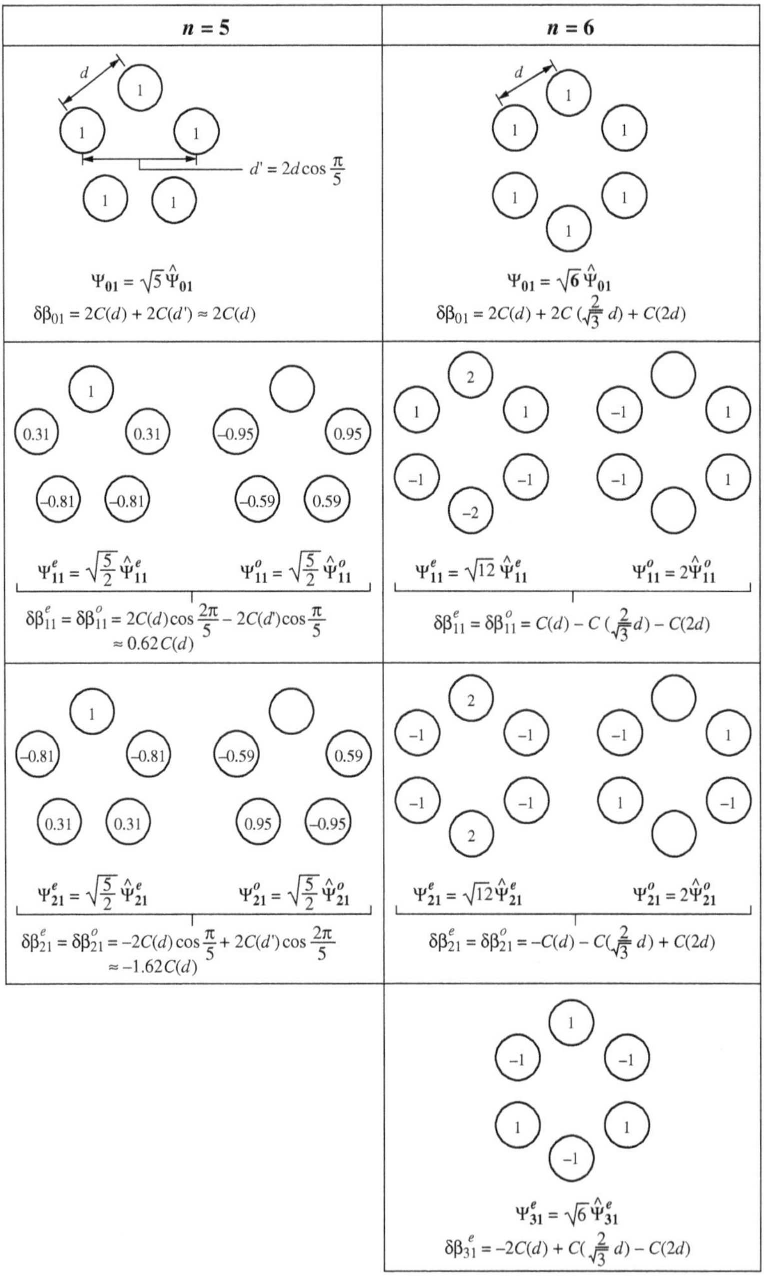

FIGURE 6.5

Three different construction procedures for vector supermodes of *n*-core fibers. In the case of supermodes that are combinations of individual core (polarization degenerate) fundamental modes, methods A and B lead to the same vector supermodes.

Competing perturbations (Sec. 6.4.3)

Δ-splitting: related to polarization component coupling (relevant for supermodes which are combinations of vector modes with two polarization components)

C-splitting: related to intercore coupling, which depends on core separation ***d*** and the individual core guidance ***V***

Core field functions for Projection Operator substitutions

Nondegenerate even and odd modes are substituted directly into Table 6.1*a* and *b*, respectively. Degenerate normalized individual core modes should be combined as $\xi_{(e)} \pm \xi_{(o)}$ and substituted into Table 6.1*c*.

Noncircular cores

For nondegenerate individual core LP modes, branch C should be used. Otherwise branches A and B can be adapted. For example, for cores with individual symmetry $\boldsymbol{C}_{lv}$, $i \geq \beta$ (e.g., see Figs. 4.5 through 4.8), for those modes that are scalar degenerate, HE/EH polarization degenerate, TE, or TM, we obtain the same mode forms as for the circular case.

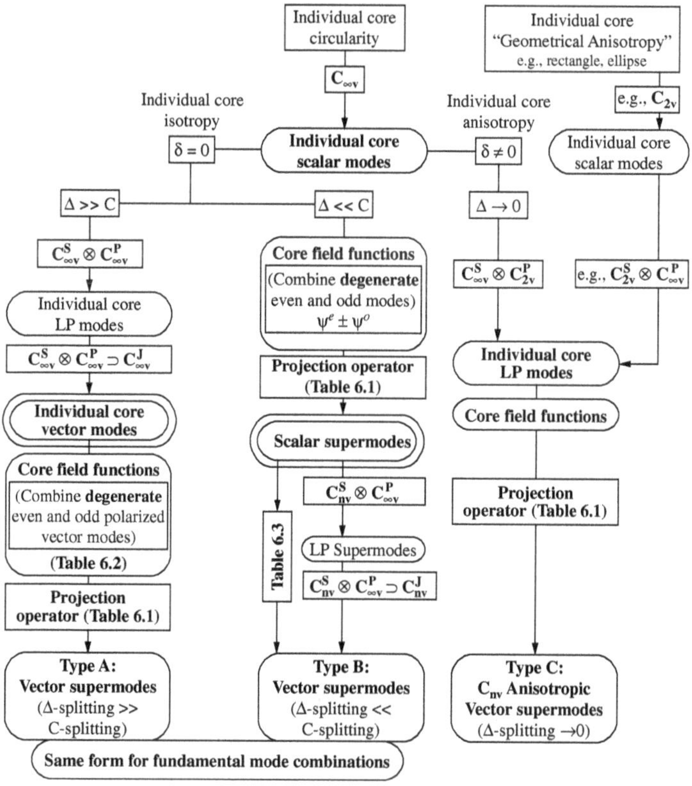

TABLE 6.3

Core Field Functions for Direct Projection Operator of Vector Supermodes

Circular Isotropic Cores
1. $\bar{\xi}_j = \bar{\xi}_j^e = \overline{TM}_j$ Nondegenerate and even $\Rightarrow$ Table 6.1*a*
2. $\bar{\xi}_j = \bar{\xi}_j^o = \overline{TE}_j$ Nondegenerate and odd $\Rightarrow$ Table 6.1*b*
3. $\bar{\xi}_j = \bar{\xi}_j^e \pm \bar{\xi}_j^o = \overline{HE}_j^e \pm \overline{HE}_j^o$ Degenerate $\Rightarrow$ Table 6.1*c*
4. $\bar{\xi}_j = \bar{\xi}_j^e \pm \bar{\xi}_j^o = \overline{EH}_j^e \pm \overline{EH}_j^o$ Degenerate $\Rightarrow$ Table 6.1*c*
Noncircular and Anisotropic Cores
They are as above except that nondegenerate hybrid or LP modes cannot be combined and thus require substitution in Table 6.1*a* or *b* depending on whether they are even or odd.
Notation
1. Overbar indicates an individual core mode with orientations $\{\hat{x}_j, \hat{y}_j\}$ as in Fig. 6.1.
2. Here *j* labels the core, and we omit subscripts denoting the individual core mode orders which are the same in any supermode combination, given identical cores.

TABLE 6.4

Weak-Guidance Construction for Vector Supermodes in Terms of the Normalized Scalar Supermodes $\hat{\psi}$ of Global Azimuthal Order L

$\hat{x}\hat{\Psi}_e^{(0)}, \hat{y}\hat{\Psi}_e^{(0)}$	$L = 0$	
$\hat{x}\hat{\Psi}_e^{(L)} \pm \hat{y}\hat{\Psi}_o^{(L)}$	$L = 1, 2, \ldots, L_n$ with	$L_n = \begin{cases} \frac{n}{2} - 1 & n \text{ even} \\ \frac{n-1}{2} & n \text{ odd} \end{cases}$
and if n is even		
$\hat{x}\hat{\Psi}_e^{(n/2)}, \hat{y}\hat{\Psi}_e^{(n/2)}$	$L = \frac{n}{2}$ *n* even	

These supermodes correspond to a $\mathbf{C}_{nv} \otimes \mathbf{C}_{\infty v} \supset \mathbf{C}_{nv}$ reduction as used in Chap. 4. The $\hat{x}$ and $\hat{y}$ polarization vectors correspond to the irrep $\mathbf{1}(\mathbf{C}_{\infty v})$ and can be taken in any direction.

same result (Sec. 6.4.2). For higher-order degenerate mode combinations, different supermode forms are obtained whose applicability depends on parameters discussed in Sec. 6.4.3. Anisotropic cores are discussed in Secs. 6.4.4 and 6.4.5.

6.4.2 Isotropic Cores: Fundamental Mode Combination Supermodes

As examples, in Fig. 6.6 we give the normalized vector supermodes for $n = 2$, 3, and 4 constructed from the fundamental modes of the individual cores. The cases $n = 5$ and 6 of course can be trivially generated in terms of the scalar supermodes of Fig. 6.4 by application of the weak-guidance reduction combinations given by Table 6.4.

These results apply to all cases of degenerate fundamental mode polarizations, i.e., isotropic cores that are circular or have individual symmetry $\mathbf{G_i} = \mathbf{C_{nv}}$ with $n \geq 3$.

Note that in application of the **projection operator construction** directly to the individual core fundamental vector modes, their degeneracy requires the use of Table 6.1*c* with the substitution of Table 6.3 core number 1 taking the explicit form

$$\bar{\xi}_j = \bar{\xi}_j^e \pm \bar{\xi}_j^o = \mathrm{HE}_{11j}^e \pm \mathrm{HE}_{11j}^o \qquad (6.14)$$

§ Orientations of the individual core modes are with respect to the n-fold symmetry planes; HE_{11j}^e and HE_{11j}^o are, respectively, polarized in the (discrete) radial and azimuthal directions with respect to the global coordinates. §

Three-Core Fiber: Isotropic

The second column of Fig. 6.6 considers the three-core fiber vector supermodes corresponding to combinations of individual fiber $lm = 01$ modes [112, 113]. Details of the constructions are given in Fig. 6.7, which illustrates the result that, for fundamental mode combinations, both the direct vector projection operator and weak-guidance constructions produce the same supermode forms. As obtained for $n = 3$ in Table 6.4, the weak-guidance constructions for these six vector supermodes are given by taking the linear combinations [112]

$$\hat{\mathbf{x}}\hat{\Psi}_{01}, \hat{\mathbf{y}}\hat{\Psi}_{01}, \hat{\mathbf{x}}\hat{\Psi}_{11}^e \pm \hat{\mathbf{y}}\hat{\Psi}_{11}^o, \hat{\mathbf{y}}\hat{\Psi}_{11}^e \pm \hat{\mathbf{x}}\hat{\Psi}_{11}^o \qquad (6.15)$$

These modes correspond to those obtained by considering the limit under which the three cores coalesce into a single circular core with the triangle perturbation symmetry maintained [112]. Under this transformation, the vector supermodes reduce to corresponding vector modes of the circular fiber; cf. the first six vector modes of Fig. 4.5.

FIGURE 6.6

Fundamental-mode combination vector supermodes for $n = 2$, 3, and 4 core isotropic fibers with global $\mathbf{C_{nv}}$ symmetry. Bracketed modes have the same propagation constants. Note that although the HE_{21}-like modes are degenerate for $n = 3$, they are split for $n = 4$. The general constructions assume weakly coupled cores immersed in an infinite cladding. However, the forms are comparable with the modes in the finite-cladding no-core limit, i.e., a single core circular fiber (Fig. 3.1) and the finite cladding $\mathbf{C_{nv}}$ perturbation or core coalescence limits (Chap. 4). Although circular cores are drawn these results also apply for cores supporting polarization degenerate fundamental modes, i.e., for individual core symmetry $\mathbf{C_{iv}}$ with $i \geq 3$.

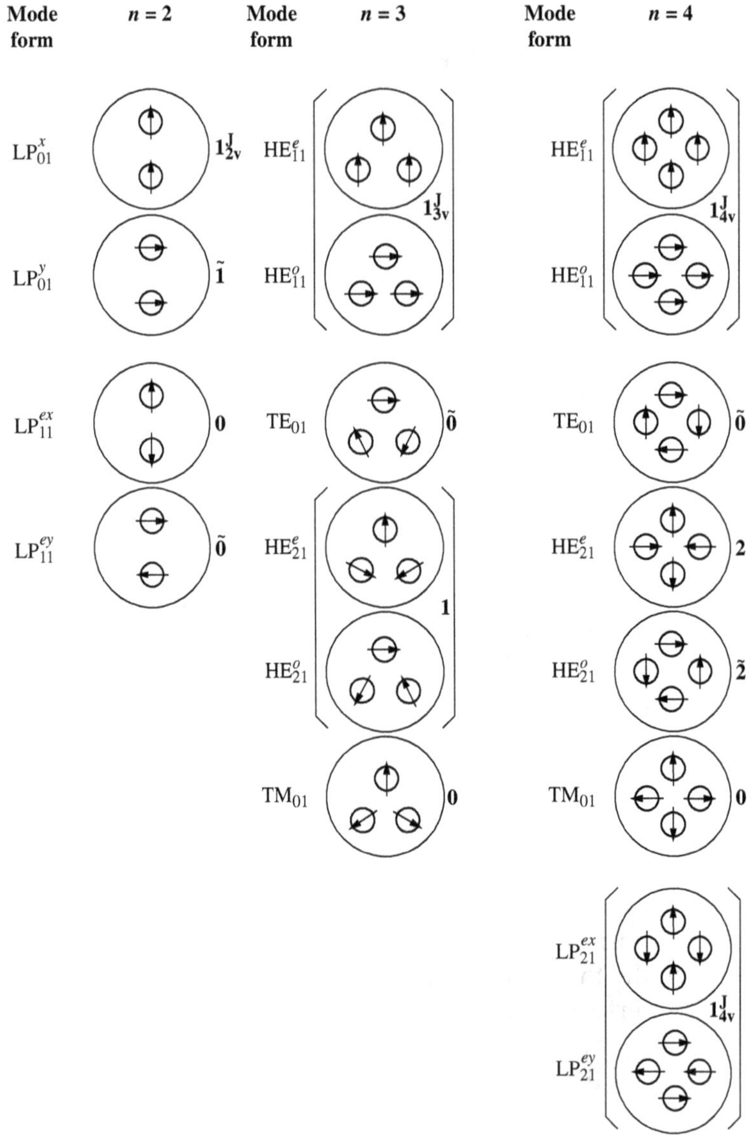

FIGURE 6.7

Illustration of the procedures outlined in Fig. 6.5 for the three-core fiber supermodes constructed from individual fiber modes with lm = 01. Note that for such fundamental mode combinations (unlike the case of degenerate higher-order mode combinations in Fig. 6.8), the general form of the vector supermodes is to lowest order independent of whether one takes into account Δ or C-splitting first.

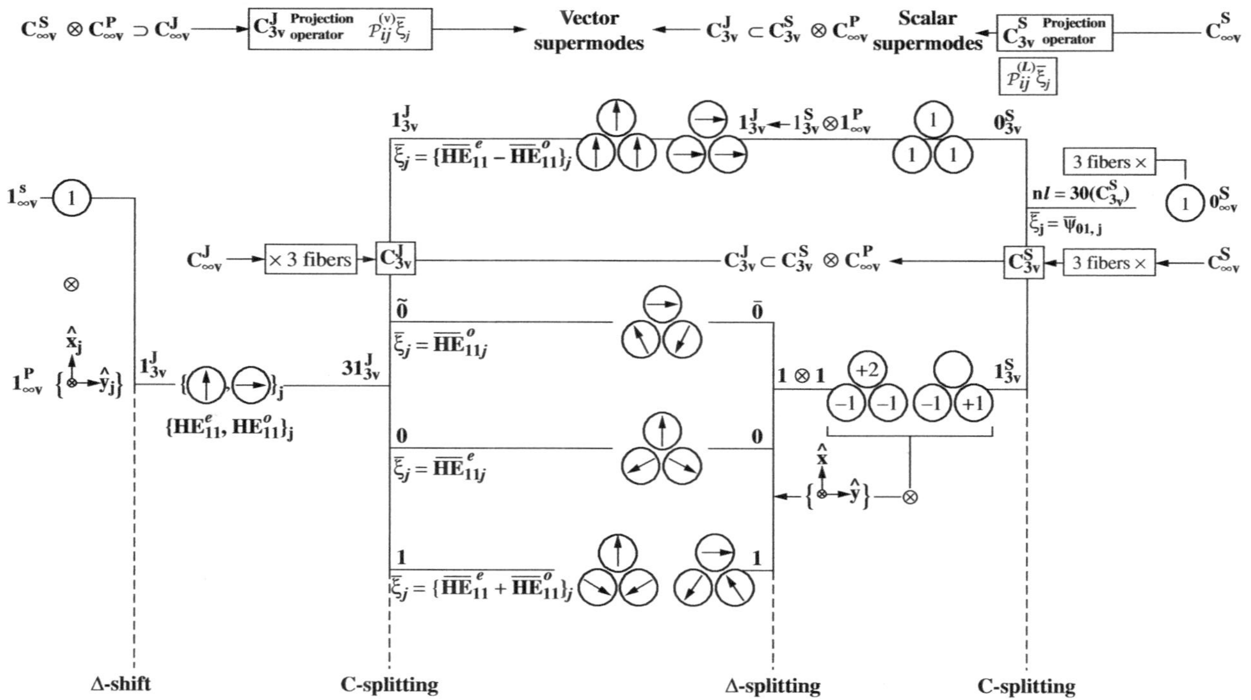

§ Note that in the $\mathbf{C_{3v}} \otimes \mathbf{C_{\infty v}} \supset \mathbf{C_{3v}}$ reduction corresponding to (6.15) the degenerate modes (the two HE pairs) transform as $\mathbf{1(C_{3v})}$, whereas the two others correspond to $\mathbf{0(C_{3v})}$ (for the even TM_{01}-like mode) and $\bar{\mathbf{0}}$ (for the odd TE_{01}-like mode). §

Four-Core Fiber: Isotropic

In the third column of Fig. 6.6 we give the vector supermodes for the four-core fiber constructed as $lm = 01$ combinations. If we consider the coalescence limit, then the four cores form a squarelike structure, and we note two marked differences between the form of the "square" supermodes and that given in the limit of coalescence to a single circular structure. (1) The circular-to-square symmetry reduction produces a splitting of the two HE_{21}-like modes. This is in contrast to the three-core case for which these modes remain degenerate. (2) The restriction of orientations means that even linearly polarized LP_{21}-like modes are supported rather than HE_{31} or EH_{11}-like modes formed on a circular structure by combinations of both even and odd LP_{21} components. Thus results are similar to those of the square perturbation; cf. the first eight vector modes of Fig. 4.6.

6.4.3 Isotropic Cores: Higher-Order Mode Combination Supermodes

For an isotropic guide, we have seen that the general form of these $l = 0$ combination supermodes holds independent of the relative magnitudes of the intercore coupling (C-splitting) and polarization-component-coupling (Δ-splitting). This is so simply because (to lowest-order weak guidance) the fundamental mode has only one polarization component.

However, for $l > 0$, this is not the case. For vector supermodes formed as combinations of higher-order modes of the individual cores, we need to consider both branches A and B in Fig. 6.5, depending on the relative magnitudes of the C- and Δ-splittings. As illustrated by the well-known example given in Fig. 18-6 of Ref. 3, depending on which effect dominates, the resulting supermodes may differ in form analogous to the examples in Chaps. 4 and 5 where we considered the competition between two perturbations such as ellipticity and finite Δ.

Polarization-Magnitude Coupling Dominates Interfiber Coupling

In the first case, when the effect of polarization-magnitude coupling due to finite Δ dominates that of interfiber coupling C, we expect the vector field in each fiber to remain approximately the same as if the fibers were isolated. The vector supermodes would thus be a linear superposition of the vector modes of the individual cores. What remains to be determined are the relative orientations and magnitudes needed to construct the supermodes. This is simply achieved by using the symmetry adapted linear combinations of Table 6.1 with core field functions being the isolated core vector modes as in Table 6.3 (see Ref. 2 for justification of combination required in cases of degeneracy).

Interfiber Coupling Dominates Polarization-Magnitude Coupling

The second case ($C >> \Delta$) consists of using the scalar supermodes as building-block structures to which polarization vectors are coupled. Here, as for the fundamental mode combinations, we use the weak-guidance construction of Table 6.4 corresponding to the group product reduction $\mathbf{C}_{nv} \otimes \mathbf{C}_{\infty v} \supset \mathbf{C}_{nv}$. The intermediate step described by product group symmetry $\mathbf{C}_{nv} \otimes \mathbf{C}_{\infty v}$ in Fig. 6.5 provides the LP supermodes of the array (by analogy with the single-core case when polarization-magnitude coupling is neglected).

Example: Two-Core Fiber: Second-Mode (lm = 11) Combinations

To illustrate the above results, in Fig. 6.8 we sketch the different steps of the constructions for the two-fiber case for supermodes corresponding to individual fiber second-mode $lm = 11$ combinations. This figure presents a qualitative understanding of the various degeneracy splittings for the two-core fiber. Analogous to the example of an elliptical fiber in Chap. 4, the vector mode transitions between the case of dominant polarization splitting and that of dominant scalar mode splitting (here intercore coupling) depend on the order of the TM_{01}/HE_{21} mode splitting, which changes at $V \approx 3.8$ for step profile cores. In particular, these are determined by the principle of *anticrossing* which forbids crossings of levels of the same symmetry labeled by the irreps $\mathbf{L}^{J}_{2v} = \mathbf{L}(\mathbf{C}^{J}_{2v})$ in Fig. 6.8, which applies to the case $V > 3.8$. See Ref. 2 for the three-core fiber case.

FIGURE 6.8

Illustration of the procedures outlined in Fig. 6.5 for the two-core fiber supermodes constructed from individual fiber modes with *lm* = 11. The form of the vector supermodes (type A or B) depends on whether the dominant effect is provided by the polarization component coupling (Δ-splitting) or intercore coupling (C-splitting).

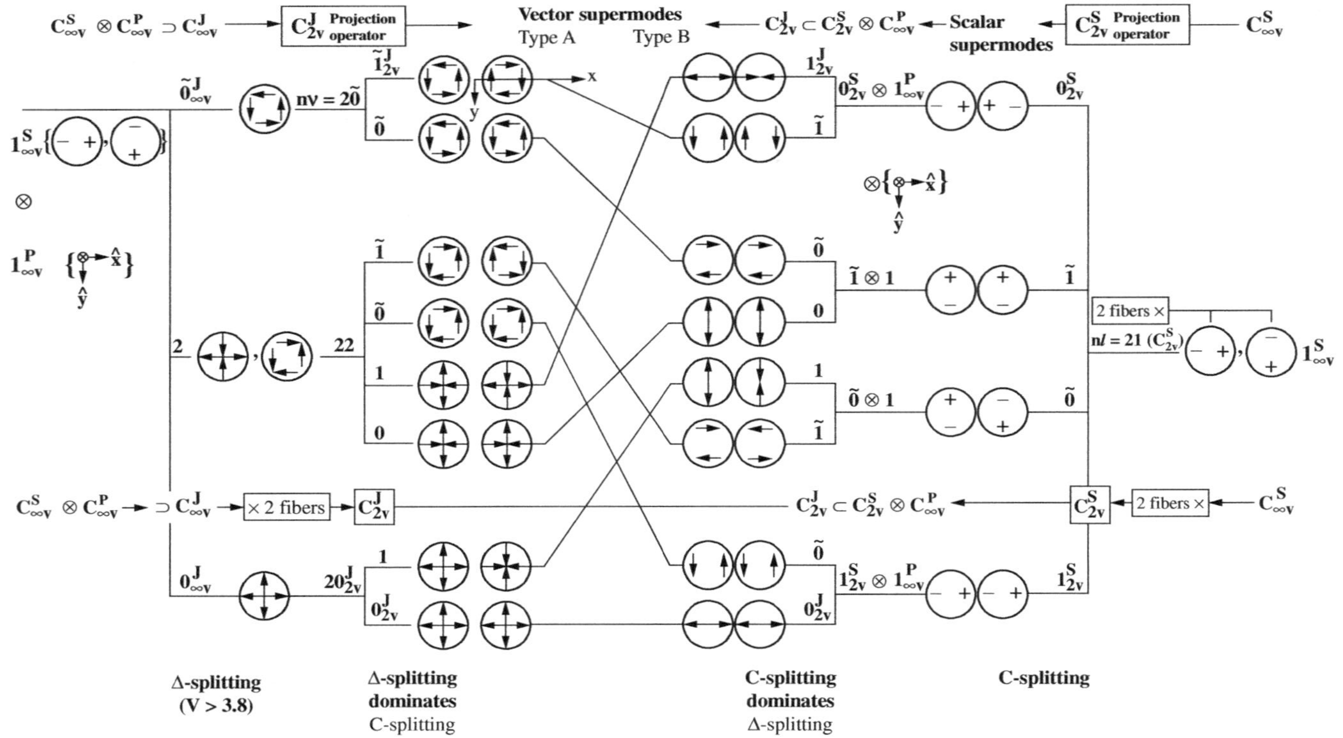

6.4.4 Anisotropic Cores: Discrete Global Radial Birefringence

Branch C of Fig. 6.5 shows how the vector supermodes may be obtained when an anisotropy compatible with the symmetry $\mathbf{C}_{\mathrm{nv}}$ is present in each core. One simply has to use Table 6.1 with

$$\begin{aligned} &\bar{\xi}_j = \bar{\xi}_j^e = \overline{\mathrm{LP}}_{01}^{xj} \qquad \text{even} \Rightarrow \text{Table } 6.1a \qquad (a) \\ &\bar{\xi}_j = \bar{\xi}_j^o = \overline{\mathrm{LP}}_{01}^{y_j} \qquad \text{odd} \Rightarrow \text{Table } 6.1b \qquad (b) \end{aligned} \tag{6.16}$$

where the overbar indicates an individual core mode with orientations $\{\hat{\mathbf{x}}_j, \hat{\mathbf{y}}_j\}$ as in Fig. 6.1.

with the degeneracy between these two orthogonal orientations in each fiber having been lifted by the anisotropy. In particular, by applying the projection operator result for $\mathbf{C}_{\mathrm{nv}}$ symmetry to the fundamental LP_{01} modes of the individual cores, we obtain the configurations in Fig. 6.9 for the case $n = 2$, 3, and 4. These mode forms may be compared with those of a radially anisotropic fiber given in the column $n \to \infty$.

Example: Three-Core Fiber: C_{3v} Anisotropy or $\Delta \to 0$ Limit The particular case $n = 3$ of Fig. 6.9 corresponds to the mode configurations shown in Fig. 2 of Ref. 114 which we refer to as *triangularly polarized* (TP) modes. TP modes provide the set of supermodes for $\mathbf{C}_{3\mathrm{v}}$ anisotropy satisfying the triangular symmetry. They are also modes of the isotropic structure in the limit $\Delta \to 0$ when all vector modes are degenerate. However, for an isotropic array with general Δ, they do not constitute a true set of vector supermodes for the structure, except for the two nondegenerate modes of TE_{01} and TM_{01}-like form. The others may be regarded as pseudo-modes somewhat analogous to LP modes. Just as the true vector supermodes of an isotropic three-core fiber (shown in Fig. 6.7) can be obtained by taking linear combinations of two LP modes, they may also be obtained by taking combinations of two of the TP modes given in Fig. 6.9.

FIGURE 6.9

Vector supermodes for n = 2, 3, and 4 cores with discrete global radial/azimuthal anisotropy in comparison with the continuum limit $(n \to \infty)$. The order of the modal propagation constants will depend on the details of structure and anisotropy. For example, a ring with index that effectively isolates the field from the central cladding will support periodic TE or TM solutions within the ring. The strong anisotropy limit of HE_{11} may be TM_{11} with a smaller propagation constant than TM_{01}.

Mode form	$n = 2$	$n = 3$	$n = 4$	Strong $n \to \infty$ radial anisotropy mode form	Weak radial anisotropy mode form
DP_{01}^{r} (LP_{11}^{ex})		TP_{01}^{r}	QP_{01}^{r}	TM_{01}	TM_{01}
$DP_{01}^{\phi 1}$ (LP_{11}^{ey})		TP_{01}^{ϕ}	QP_{01}^{ϕ}	TE_{01}	TE_{01}
DP_{11}^{er} (LP_{01}^{x})		TP_{11}^{er}	QP_{11}^{er}	TM_{11}^{e}	HE_{11}^{e}
$DP_{11}^{o\phi}$ (LP_{01}^{y})		TP_{11}^{or}	QP_{11}^{or}	TM_{11}^{o}	HE_{11}^{o}
		$TP_{11}^{e\phi}$	$QP_{11}^{e\phi}$	TE_{11}^{e}	EH_{11}^{o}
		$TP_{11}^{o\phi}$	$QP_{11}^{o\phi}$	TE_{11}^{o}	EH_{11}^{e}
			QP_{21}^{er}	TM_{21}^{e}	HE_{21}^{e}
			$QP_{21}^{e\phi}$	TE_{21}^{e}	EH_{21}^{o}

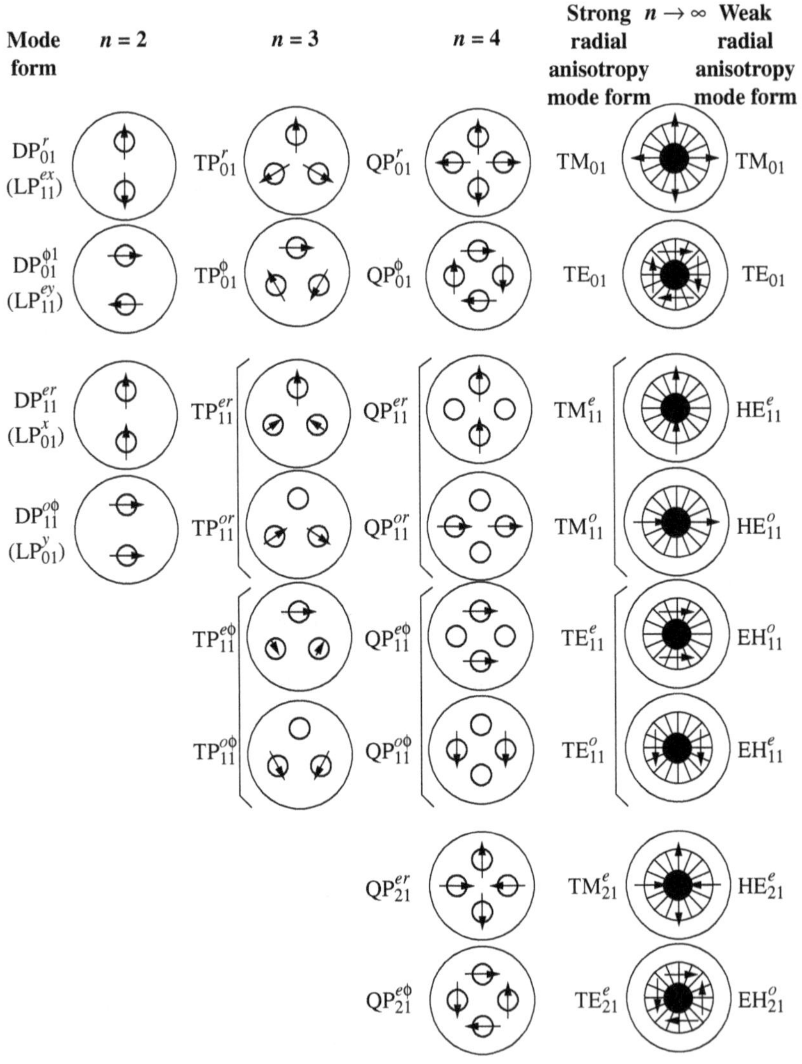

6.4.5 Other Anisotropic Structures: Global Linear and Circular Birefringence

§ For a multicore fiber with global linear birefringence, i.e., with birefringent axes independent of position and in the directions $\hat{\mathbf{x}}$ and $\hat{\mathbf{y}}$, for example, the true weak-guidance vector supermodes are simply the LP supermodes formed as products of the scalar supermodes and $\hat{\mathbf{x}}$ or $\hat{\mathbf{y}}$. The symmetry is $\mathbf{C}^{\mathrm{S}}_{\mathrm{nv}} \otimes \mathbf{C}^{\mathrm{P}}_{2\mathrm{v}}$, that is, the discrete analog of the $\mathbf{C}^{\mathrm{S}}_{\infty\mathrm{v}} \otimes \mathbf{C}^{\mathrm{P}}_{2\mathrm{v}}$ symmetry for linearly birefringent single circular core fibers in Sec. 5.1. Propagation constant splitting due to finite Δ is then given by the reduction to joint symmetry $\mathbf{C}^{\mathrm{S}}_{\mathrm{nv}} \otimes \mathbf{C}^{\mathrm{P}}_{2\mathrm{v}} \supset \mathbf{C}^{\mathrm{J}}_{2\mathrm{v}}$, also analogous to that in Sec. 5.1.

Similarly, for global circular birefringence, the true supermodes are CP (except for the TM_{0m} and TE_{0m}-like supermodes). The appropriate symmetry reduction is $\mathbf{C}^{\mathrm{S}}_{\mathrm{nv}} \otimes \mathbf{C}^{\mathrm{P}}_{\infty} \supset \mathbf{C}^{\mathrm{J}}_{\mathrm{n}}$, that is, the discrete analog of the circular single-core result in Sec. 5.3. The associated irrep branching rules are $l(\mathbf{C}^{\mathrm{S}}_{\mathrm{nv}}) \otimes \pm 1(\mathbf{C}^{\mathrm{P}}_{\infty}) \rightarrow +\nu(\mathbf{C}^{\mathrm{J}}_{\infty})$; cf. Fig. 5.5. §

6.5 GENERAL NUMERICAL SOLUTIONS AND FIELD APPROXIMATION IMPROVEMENTS

6.5.1 SALCs as Basis Functions in General Expansion

In general, given a set of core field functions $\bar{\xi}_j$, the symmetry adapted linear combinations $\xi_i^{(L)}$ generated therefrom simply provide a set of basis functions for construction of supermode fields. For example, for the even scalar supermodes of $\mathbf{C}_{\mathrm{nv}}$ structures, we should consider all even SALCs of (global) azimuthal symmetry L [(Chap. 13)] constructed from all the modes of the individual cores, i.e.,

$$\Psi_e^{(L)} = \sum_{ilm} a_{ei}(lm)\xi_{ei}^{(L)}(lm) \tag{6.17}$$

where the coefficients a_{ei} in general must be determined numerically, e.g., either directly via matching the boundary conditions or via a variational approach [115]. (Strictly speaking, the summation should also include an integral over the continuous spectrum of radiation mode SALCs.)

6.5.2 Variational Approach

For the scalar supermodes Ψ, from the SWE (2.15) in the notation of Eq. (2.22) (and suppressing the tilde for brevity), we obtain, after multiplication by Ψ and integration over the infinite cross section,

$$\beta^2 = <\psi, \mathcal{H}_S \psi> / <\psi, \psi> \tag{6.18a}$$

This may be used variationally by substitution of the SALC expansion (6.17) and the standard Rayleigh-Ritz procedure of varying the coefficients a_i so as to maximize β, that is, by solving the set of equations

$$\frac{\partial \beta^2}{\partial a_i} = 0 \tag{6.18b}$$

6.5.3 Approximate SALC Expansions

Fundamental Mode Combination Supermodes

As a first approximation, in Sec. 6.3 we obtained the fundamental mode combination supermodes as given directly by the fundamental mode SALCs

$$\Psi_e^{(L)} = \xi_e^{(L)}(01) \tag{6.19}$$

These supermode field forms are independent of intercore separation d.

This approximation may be improved to take account of d by allowing a contribution to the field from, e.g., the second-mode SALCs, i.e., those constructed from the isolated core modes with $lm = 11$ (assuming they are bound modes). For example, substitution in Eq. (6.18) of the trial functions

$$\Psi_e^{(L)} = \begin{cases} \xi_e^{(L)}(01) + a\xi_e^{(L)}(11) & L = 0, \frac{n}{2} \quad (6.20a) \\ \xi_e^{(L)}(01) + a_1\xi_{e1}^{(L)}(11) + a_2\xi_{e2}^{(L)}(11) & 0 < L < \frac{n}{2} \quad (6.20b) \end{cases}$$

leads to one- and two-parameter variational problems, respectively. Obviously the variational coefficients of Eqs. (6.20) quickly tend to zero with increasing separation as the simplest fundamental mode combination supermode form of Eq. (6.19) is known (especially in the two-core case) to be an excellent approximation even when the cores are quite close.

Second-Mode Combination Supermodes

Consider the case $0 < L < n/2$. If we assume the supermode form

$$\Psi_e^{(L)} = a_1\xi_{e1}^{(L)}(11) + a_2\xi_{e2}^{(L)}(11) \qquad (6.21)$$

then substitution into Eq. (6.18) and invoking the hermitian nature of H_S lead to the requirement [2] that $a_1/a_2 = \pm 1$, that is,

$$\Psi_e^{(L)} = \xi_{e\pm}^{(L)}(11) = \xi_{e1}^{(L)}(11) \pm \xi_{e2}^{(L)}(11) \qquad (6.22)$$

This supermode field form in terms of the two SALCs of the same global parity is the first approximation also obtained via the recipe of Sec. 6.3.3 [i.e., substitution into Table 6.1*c* of a core field function (6.13) formed by combining the two degenerate individual core modes (of opposite parity)]. To improve the approximation, we could take

$$\Psi_e^{(L)} = a_0\xi_e^{(L)}(01) + a_1\xi_{e1}^{(L)}(11) + a_2\xi_{e2}^{(L)}(11) \qquad (6.23)$$

which is the same as Eq. (6.20*b*) except that we now use the second-mode SALC $\Psi_e^{(L)} = \xi_{e\pm}^{(L)}(11)$ as the starting solution. Choosing $a_1 = 1$, this is conveniently written as the two-parameter trial function

$$\Psi_e^{(L)} = \xi_{e\pm}^{(L)}(11) + a_0\xi_e^{(L)}(01) + \delta a_2\xi_{e2}^{(L)}(11) \qquad (6.24)$$

6.5.4 SALC = Supermode Field with Numerical Evaluation of Sector Field Function

As an alternative to the SALC expansions, one may pose the general numerical problem as follows: Symmetry requires that for some $\bar{\xi}_j$ the supermode field be of a form given by the SALCs of Table 6.1. This can be used to reduce the computation required in a general numerical method involving e.g., finite elements or finite differences. In fact, for harmonic expansions of the field, the general numerical problem reduces to satisfying the boundary conditions on just one-half of the n cores.

Note that although, for a given supermode, determination of $\bar{\xi}_j$ for only one j is required (as the others are given directly by rotation), for each different supermode $\bar{\xi}_j$ is different except for the degenerate supermode pairs. However, as a starting solution for general numerical evaluation, of course one can make use of the fact that to a first approximation there is a set of supermodes with $\bar{\xi}_j \approx$ the same isolated core mode, i.e., the approximate forms considered in Sec. 6.3.

6.5.5 Harmonic Expansions for Step Profile Cores

For step profile cores we may expand the even and odd parts of the scalar field associated with the jth core, $\bar{\xi}_j^e$ and $\bar{\xi}_j^o$, in the jth core as

$$\bar{\xi}_j^e = \sum_{l=1}^{L} a_{el} J_l(UR_j)\cos l\phi_j \qquad \xi_j^0 = \sum_{l=1}^{L} a_{ol} J_l(UR_j)\sin l\phi_j \qquad R_j \equiv \frac{r_j}{\rho} \tag{6.25a}$$

and in the cladding as

$$\bar{\xi}_j^e = \sum_{l=1}^{L} b_{el} K_l(WR_j)\cos l\phi_j \qquad \xi_j^o = \sum_{l=1}^{L} b_{ol} K_l(WR_j)\sin l\phi_j \tag{6.25b}$$

In general the coefficients must be determined numerically, e.g., using two boundary conditions, namely, continuity of the scalar field ψ and its first derivative ψ'.

For couplers usually this has been done directly for the full vector modes [71, 116–118] after (1) expanding for the longitudinal components of the electric and magnetic fields as in Eq. (6.13), (2) obtaining the transverse components in terms of the longitudinal components (via the explicit relations [3 (Eq. 30-5)]), and (3) matching four boundary conditions: continuity of the tangential field components and normal displacement field components, i.e., for core i, $\mathbf{e}\cdot\hat{\mathbf{t}}_i$, $\mathbf{h}\cdot\hat{\mathbf{t}}_i$, $n^2\mathbf{e}\cdot\hat{\mathbf{n}}_i$ and $n^2\mathbf{h}\cdot\hat{\mathbf{n}}_i$ where the tangential and normal unit vectors reduce to $\hat{\phi}_i$ and $\hat{\mathbf{r}}_i$ for circular cores.

However, it may often be advantageous to obtain the scalar field numerically and then determine the vector mode properties by using the perturbation approach here.

For step-core fibers, two methods are used for determining the expansion coefficients together with the eigenvalue U (and thus W). Although they have mostly been applied to the vector modes, they are straightforwardly adapted for the scalar modes with simplification. These are

1. *Point-matching method* [71, 116], i.e., direct substitution of the circular harmonic expansions and matching of the field and its derivative at chosen points on a core-cladding boundary;
2. *Addition formula expansion* of $K_l(WR_j)$ at the boundary of the ith core in terms of R_i [117, 118]. As matching occurs at $R_i = 1$, it is no longer necessary to choose arbitrary points for the matching. However, the algebraic complexity increases because of the extra summation required for reexpression in terms of R_i.

We also refer to the methods discussed in Refs. 99 and 119, which have been applied to single-core guides but have potential for application to coupled structures.

6.5.6 Example of Physical Interpretation of Harmonic Expansion for the Supermodes

To obtain some physical insight into the meaning of the above expansions, let us consider the effect of core separation on the supermodes with approximate constructions in terms of the individual core fundamental modes. To zeroth order in the core separation, we have simply

$$\bar{\xi}_j^e = \bar{\psi}_{01}(UR_j), \qquad \bar{\xi}_j^o = 0 \qquad \text{thus} \qquad \bar{\xi}_j = \bar{\xi}_j^e + \bar{\xi}_j^o = \bar{\psi}_{01}(UR_j) \tag{6.26}$$

To take account of finite core separation, let us suppose a situation in which the terms of azimuthal order $l = 1$ provide the dominant correction, i.e.,

$$\bar{\xi}_j^e = \bar{\psi}_{01}(UR_j)_j + a_{e1}J_1(UR_j)\cos\phi_j \; \xi_j^o = a_{01}J_1(UR_j)\sin\phi_j \tag{6.27}$$

$$\bar{\xi}_j = \bar{\xi}_j^e + \bar{\xi}_j^o = \bar{\psi}_{01}(UR_j) + \mathrm{J}_1(UR_j)(a_{e1}\cos\phi_j + a_{01}\sin\phi_j) \tag{6.28}$$

In providing a physical interpretation it is convenient to reexpress this as

$$\xi_j = \bar{\psi}_{01}(UR_j) + a_1 J_1(UR_j)\cos(\phi_j - \phi_{\text{tilt}}) \tag{6.29}$$

where $U = \bar{U} + \delta U$ (see Sec. 6.5.7), $a_1 = \sqrt{a_{e1}^2 + a_{o1}^2}$, and $\phi_{\text{tilt}} = \arctan(a_{o1}/a_{e1})$ depend on core separation d and are determined via the boundary conditions. Thus the effect of considering finite separation d for the field shape in a given core is a bulge in the direction ϕ_{tilt} with respect to the core symmetry axis. In the case of the fundamental supermode (for which the sector field function $\bar{\psi}_j$ is even, that is, $a_{oi} = 0$), $\phi_{\text{tilt}} = 0$ and bulge is simply toward the center of the multicore fiber. However, for the supermodes corresponding to $1 \leq L < n/2$, we expect ϕ_{tilt} for the sector field function to be nonzero.

6.5.7 Modal Expansions

In the above harmonic expansions, the eigenvalue U is the same for each term $J_l(UR_j)$ and corresponds to the supermode eigenvalue being varied until the boundary conditions are matched. In a first approximation $U \approx \bar{U}$ = eigenvalue of the isolated core mode from which the supermode is constructed in a first approximation.

By contrast in a modal expansion, supposing for simplicity that the second-order modes are bound, we have

$$\bar{\psi}_j = \bar{\psi}_{01}(U_{01}R_j) + a_1 J_1(U_{11}R_j)\cos(\phi_j - \phi_{\text{tilt}}) \tag{6.30}$$

where U_{01} etc. are the eigenvalues of the modes in isolation.

6.5.8 Relation of Modal and Harmonic Expansions to SALC Expansions

§ Expansion of the core field function in such a form is effectively the same as the SALC expansions considered earlier, but provides some additional insight regarding the field form. Note that the harmonic expansions are also effective SALC expansions, as is seen by the summation inversion convenient for numerical evaluation [71]. The harmonic expansion has the attraction of allowing treatment of single-mode cores without the necessity of dealing with radiation modes. §

6.5.9 Finite Claddings and Cladding Modes

§ The latter radiation problem is sometimes avoided in the practical case of a finite cladding at which the external interface field is conveniently set equal to zero and the continuum of radiation modes of an infinite-cladding guide becomes the discrete spectrum of cladding modes. These have received considerable study in the case of tapered fibers [31, 32, 39, 41, 120]. For supermodes near *cutoff* or, in the case of finite-cladding fibers, what is known as core-mode cutoff [97] (or the *core-clad transition*), the external interface can play a crucial role in determination of the field form. §

6.6 PROPAGATION CONSTANT SPLITTING: QUANTIFICATION

In this section we quantify supermode propagation constant splitting for multicore fibers of $\mathbf{C}_{nv}$ symmetry using the results for two-core fibers as building blocks to obtain for each supermode a *correction* to the isolated core propagation constant. Splitting is of interest for determination of beat lengths and thus power transfer characteristics as in Sec. 6.7.

6.6.1 Scalar Supermode Propagation Constant Corrections

The scalar supermode propagation constant β is related to that of the propagation constant $\bar{\beta}$ of the mode of an isolated core by the reciprocity relation of Eq. (2.28) in the form

$$\beta^2 - \bar{\beta}^2 = k^2 \frac{\int_{A_\infty} (n^2 - \bar{n}_i^2)\Psi\bar{\psi}_i dA}{\int_{A_\infty} \Psi\bar{\psi}_i dA} \tag{6.31}$$

where $\bar{n}_i$ and $\bar{\psi}_i \neq 0$ are the refractive index profile and scalar mode of the reference core i in isolation, and n and Ψ are the profile and scalar supermode of the composite structure. This expression for the scalar supermode propagation constant is exact given that we know Ψ exactly. However, for practical evaluation several approximations are usually made:

1. Within a perturbation framework, we take the supermodes to be given by the symmetry construction of previous

sections whose linear combination form is independent of intercore separation d, that is,

$$\Psi = \sum_{j=1}^{n} a_j \bar{\psi}_j + O\left(\frac{1}{d}\right) \approx \sum_{j=1}^{n} a_j \bar{\psi}_j \tag{6.32}$$

2. We assume that the cores are electromagnetically well separated: $\bar{\psi}_j (j \neq i)$, which decreases exponentially with distance from core j, is neglected in fiber i, and thus only the product $\bar{\psi}_i \bar{\psi}_i$ contributes in the denominator (there is a wealth of recent literature including corrections to this approximation for the two-core case).
3. Finally, small index differences so that $\delta\beta \equiv \beta - \bar{\beta} \approx (\beta^2 - \bar{\beta}^2)/2kn$ etc. lead to

$$\delta\beta = k \sum_{j \neq i} a_j \frac{\int_{A_\infty} (n - \bar{n}_i) \bar{\psi}_i \bar{\psi}_j \, dA}{a_i \int_{A_\infty} \bar{\psi}_i^2 \, dA} \tag{6.33}$$

Fundamental Mode Coupling

In the case of n cores distributed with $\mathbf{C}_{\mathbf{nv}}$ symmetry, for the supermodes constructed from the azimuthally independent $l = 0$ modes of the individual cores, the relative amplitude of the field in the cores is given by $a_j = a_i \cos[2\pi L(j - i)/n]$. Thus by defining

$$C(d_{ij}) = k \frac{\int_{A_\infty} (n - \bar{n}_i) \bar{\psi}_i \bar{\psi}_j \, dA}{\int_{A_\infty} \bar{\psi}_i^2 \, dA} \tag{6.34}$$

as the usual (weak) coupling coefficient for two cores or fibers (i and j) which have centers separated by distance d_{ij}, noting that for the multicore array this distance is related to the adjacent core separation d by

$$d_{ij} = \frac{d \sin[\pi(i - i)/n]}{\sin(\pi/n)} \tag{6.35}$$

and, for convenience, taking core n as the reference, Eq. (6.34) reduces to

$$\boxed{\delta\beta_L = \sum_{j=1}^{n-1} C(d_{jn}) \cos\left(\frac{2\pi L j}{n}\right)} \tag{6.36}$$

For $n = 2$ cores of center separation d, this gives the usual result

$$\delta\beta_0 = C(d) \qquad \text{and} \qquad \delta\beta_1 = -C(d) \tag{6.37}$$

For $n \geq 3$, considering *nearest-neighbor* (nn) interaction only, i.e., between adjacent cores of separation d, Eq. (6.36) reduces to [73]

$$\boxed{\delta\beta_L \approx C(d)\cos\left(\frac{2\pi L}{n}\right)} \tag{6.38}$$

Particular cases for n and L are given in Table 6.5; see also Figs. 6.3 and 6.4 for the cases $n \leq 6$.

TABLE 6.5

Scalar Supermode Propagation Constant Corrections and Corresponding Vector Supermode Degeneracies for a C_{nv} Array of n Weakly Coupled Cores in an Infinite Cladding

Mode Form	Two Cores	Three Cores	Four Cores	*n* Cores
L = 0	$C(d)$	$2C(d)$	$2C(d) + C(\sqrt{2}\,d)$	$2C(d)$
LP^x_{01} HE^e_{11} LP^y_{01} HE^o_{11}	Nondegenerate	Degenerate	Degenerate	Degenerate
L = 1	$-C(d)$	$-C(d)$	$0 - C(\sqrt{2}\,d)$	$2\cos(2\pi/n) \times C(d) - \ldots$
LP^{ex}_{11} LP^{ey}_{11}	Nondegenerate	—	—	
TE_{01} TM_{01} HE^e_{21} HE^o_{21}		Degenerate	Nondegenerate	Degenerate
L = 2		—	$-2C(d) + C(\sqrt{2}\,d)$	$2\cos(4\pi/n) \times C(d) + \ldots$
LP^{ey}_{21} LP^{ey}_{21}		—	Degenerate	

Scalar mode corrections are in terms of corrections. $\delta\beta_{scalar} = \pm C(d)$ for a two-core system, where d = separation of two nearest-neighbor cores = $2r_2 = \sqrt{3}\,r_3 = \sqrt{2}\,r_4$, and r_n = radius of the circle on which the n cores are symmetrically distributed.

Note that for step profiles, $C(d)$ is given, in terms of the notation of Table 2.1, by [3]

$$C(d) = \frac{\sqrt{2\Delta}U^2}{\rho V^3} \frac{K_0(Wd/\rho)}{K_1^2(W)} \tag{6.39}$$

§ Second-Mode Coupling

As a second example, we note that for the case $n = 2$ and $l = 1$ for which the corresponding four non-degenerate modes are given in Fig. 6.8, the propagation constants are given by $\delta\beta_{e\pm} = \pm C_+$, and $\delta\beta_{o\pm} = \pm C_-$, where even/odd modes are defined with respect to the interfiber (x) axis and, in the notation of Table 2.1 and [3],

$$C_\pm = \sqrt{\frac{\Delta}{2}} \frac{U^2}{\rho V^3} \frac{K_2(Wd/\rho) \pm K_0(Wd/\rho)}{K_0(W)K_2(W)} \tag{6.40}$$

Equation (6.40) can also serve as a building block for structures with more than two cores, although one now needs to account for the different orientations of the azimuthal dependencies of the fields in each fiber. §

6.6.2 Vector Supermode Propagation Constant Corrections

§ For the vector mode problem we have two "perturbation" parameters, Δ and d. Thus comments analogous to those in Sec. 4.5 for elliptical guides apply. Given a sufficiently accurate scalar field, determined to take account of fiber separation (e.g., using either a general numerical or an approximate variational approach as described in Sec. 6.5), the lowest-order weak-guidance vector field can be generated using the combinations of Sec. 6.4. Substitution into Eq. (2.21) then gives the first-order vector corrections to first order in Δ. In both numerical and analytic evaluation of Eq. (2.21), symmetry can provide considerable simplification, particularly for the multicore problem.

Depending on the type of structure and index differences under consideration, numerical determination of the scalar supermodes followed by numerical evaluation of the polarization corrections using the integral expression (2.21) [121] may often be preferable to direct numerical solution of the vector eigenvalue problem. As well as producing computational savings, this approach can reduce the problem of susceptibility to rounding error particularly when determining small splittings in the supermode propagation constants.

We remark that the vector corrections can be quite sensitive to small imperfections such as nonuniform stress, core placement, and core profile variations (e.g., the supermodes are highly sensitive to nonuniformity in the fiber separation, and we saw in Sec. 3.4 that profile grading had a considerable effect). Thus their calculation for ideal step cores is often more of academic interest. §

6.7 POWER TRANSFER CHARACTERISTICS

From the practical viewpoint, the major interest in multicore fibers (or equivalently multifiber couplers) is the intercore or interfiber power transfer. This may be analyzed in terms of coupling between individual fiber modes as in [42, 122, 123] or supermode beating [3] as here.

The two types of structure with symmetry that have attracted particular attention to date are (1) the $n \times n$ coupler and (2) the $1 \times n$ coupler. The former simply corresponds to the n-core fiber with n cores (or fibers) arranged in a ring, i.e., the structure analyzed in previous sections. The latter is usually simply a generalization of this which retains discrete azimuthal symmetry, i.e., consists of central core (or fiber) surrounded by $n-1$ cores arranged in a ring with $\mathbf{C}_{\mathbf{nv}}$ symmetry. However, sometimes the symmetry is lowered by placing the guides in the ring arrangement in pairs; e.g., Mortimore considers a structure where the symmetry of a 1×5 multifiber coupler (i.e., one central fiber surrounded by four fibers in the ring) is lowered from $\mathbf{C}_{\mathbf{4v}}$ to $\mathbf{C}_{\mathbf{2v}}$.

We concentrate on type (1) structures and mention the generalization to type (2) structures, referring to [73, 122] for further results.

6.7.1 Scalar Supermode Beating

Type (1) Structures Given excitation in core i of isolated core mode $\bar{\psi}_i$, after a length L, the *power* in isolated core mode $\bar{\psi}_j$ is given in terms of a superposition of scalar supermodes ψ_m via the modal interometric formula [124]

$$I_j = \sum_m I_m + 2\sum_{m>n}\sum \sqrt{I_m I_n}\cos\Phi_{mn} \tag{6.41}$$

where

$$\Phi_{mn} = (\beta_m - \beta_n)L = \text{phase difference between supermodes } m \text{ and } n \text{ at } z = L \quad (6.42a)$$

$$I_m = |a_{m0}a_{mL}|^2 = \text{supermode powers} \quad (6.42b)$$

with

$$a_{m0} = <\bar{\psi}_i, \psi_m> = \text{overlap of input coremode } i \text{ with supermode } m$$

$$a_{mL} = <\psi_m, \bar{\psi}_j> = \text{overlap of input coremode } i \text{ with supermode } m$$

In Table 6.6, we give the scalar supermode combinations corresponding to the excitation of a single core.

For the cases $n = 2$ and 3, this is simply achieved with just two supermodes, resulting in a simple sinusoidal power transfer to and from the excited fiber. (For ideal $n = 2$ and $n = 3$ couplers, 100 percent and eight-ninths of the power is transferred from the excited fiber.) However, for $n = 4$ and 5, three supermodes are required; for $n = 6$, 7 four are required; etc. Nevertheless in the case $n = 4$, the propagation constant splittings are approximately equal, simplifying the power transfer characteristics.

For a full understanding of the power transfer particularly for strongly fused couplers, it is interesting to compare the given supermode fields and propagation constant splittings with the corresponding modes and splittings of both a circular fiber and an n-sided polygon.

Type (2) Structures Symmetry simply restricts the form of supermodes of the type (2) structures to be $a_o\psi_{lm} \pm a_{\text{ring}}\Psi^{(l)}$. In general, a_o and a_{ring} are not restricted by symmetry and must be determined numerically. However, for the majority of modes either a_o or a_{ring} will be negligible. Thus for power transfer between the central core and those of the ring, one is only interested in two modes corresponding to the central and ring modes of the same symmetry and equal propagation constants, for example, $a_o\psi_{01} \pm a_{\text{ring}}\Psi_{01}$. As we have seen that Ψ_{01} will have its propagation constant increased with respect to that of the individual cores in the ring, for ideal phase matching with the central core, one could raise its propagation constant by a slight increase in either its size or its index with respect to the other cores.

TABLE 6.6

Power Transfer Characteristics of n-Core Fibers Given Excitation of one Core within the Scalar Mode and Weak-Coupling Approximations

n	Supermode Combination Corresponding to Single-Fiber Excitation	Propagation Constant Splitting $\delta\beta_{ps\text{-}qt} \equiv \beta^s_{p1} - \beta^t_{p1} = \delta\beta_{nn} + \delta\beta_w$ (nn = nearest neighbor, w = weak): Supermode Pair	$C = C(d)$: $\delta\beta_{nn}$	$\delta\beta_w$	N_{bf} (nn): nn	N_{bf} (oc): + Weak Terms	N_{bf} (am): All Modes
2	$\Psi_{01} + \Psi_{11}$	0-1	$2C$		1	1	1
3	$\Psi_{01} + \Psi^e_{11}$	0-1e	$3C$	0	1	1	3
4	$\Psi_{01} + 2\Psi^e_{11} + \Psi^e_{21}$	0-1e	$2C$	$+2C_w$	2	3	6
		1e-2e	$2C$	$-2C_w$			
	$C_w = C(\sqrt{2}d)$	0-2e	$4C$	0			
5	$\Psi_{01} + 2\Psi^e_{11} + \Psi^e_{21}$	0-1e	$1.382C$		3	3	10
	$C_w = C(d') \approx C(1.62d)$	1e-2e	$2.236C$				
	$d = 2d\cos(\pi/5) \approx 1.62d$	0-2e	$3.618C$				
6	$2\Psi_{01} - \Psi^e_{11} + \Psi^e_{21} - 2\Psi^e_{31}$	0-1e	C	$+3C_w + 2C_{vw}$	4	6	15
		2e-3e	C	$-2C_w + 2C_{vw}$			
	$C_w = C\left(\frac{2}{\sqrt{3}}\right)d$	1e-2e	$2C$	$0 + 2C_{vw}$			
		0-2e	$3C$	$+3C_w$			
	$C_{vw} = C(2d)$	1e-3e	$3C$	$-2C_w$			
		0-3e	$4C$	$+C_w + 2C_{vw}$			

$\delta\beta_{nn}$ is due to nearest-neighbor coupling; $\delta\beta_w$ takes account of the other cores. The superposition is in terms of the unnormalized supermodes of Figs. 6.3 and 6.4.

6.7.2 Polarization Rotation

The general feature, similar to the case of the $lm = 11$ modes of an ideal single-core circular fiber as in Sec. 3.3.4, is that a linearly polarized supermode excitation usually corresponds to excitation of two true vector supermodes which in general beat with a consequential rotation of the polarization.

For supermodes that formed as different combinations of the same vector mode of the individual cores, the polarization splittings between these supermodes are of order Δ smaller than the crosstalk splittings discussed above. Thus the details can be quite sensitive to nonideal situations (e.g., small residual birefringence, distortions of the circular symmetry of each fiber core, distortions from $\mathbf{C}_{\mathbf{nv}}$ array symmetry), and details of the predicted consequences are of greater academic interest. Furthermore, although the scalar mode splittings provide a useful first approximation, the additional polarization splittings are subject to considerable approximation. The feature of particular interest is qualitative prediction of splitting or degeneracy for the ideal structure. For near-ideal structures this allows identification of the number of different dominant beat lengths for a given excitation.

In Table 6.7 we consider the vector supermode combinations for n-core fibers corresponding to excitation of a single core in the global radial ($\hat{\mathbf{x}}_j$) or azimuthal ($\hat{\mathbf{y}}_j$) direction.

§ Consider in detail the case of $n = 3$. The dominant beating will be between two LP supermodes (LP_{01} and LP_{11}). For isotropic circular cores LP_{11} is only a *pseudo*-supermode and corresponds to two *true* supermodes (e.g., $LP_{11}^{ex} = HE_{21}^{e} + TM_{01}$). Thus dominant beating will be modulated by the polarization beating of these true supermodes with a consequential polarization rotation, for example, LP_{11}^{ex} to LP_{11}^{oy} as in Sec. 3.3.4. However, if the three-core fiber is globally, linearly, or radially anisotropic, then the beating

TABLE 6.7

Vector Supermode Combinations Corresponding to Polarized Single-Fiber Excitation of the Upper (nth) Fiber

n	Single-Core Excitation	Scalar Supermode Construction	LP Supermode Combination	True Vector Supermode Combination: Isotropic Cores	Supermode Combination for C_{nv} Anisotropy
2	$\overline{LP}_{01}^{x_2}$	$\{\Psi_{01} + \Psi_{11}\}\hat{\mathbf{x}}$	$LP_{01} + LP_{11}^{x}$	$LP_{01} + LP_{11}^{x}$	$DP_{11}^{r} + DP_{01}^{r}$
	$\overline{LP}_{01}^{y_2}$	$\{\Psi_{01} + \Psi_{11}\}\hat{\mathbf{y}}$	$LP_{01} + LP_{11}^{y}$	$LP_{01} + LP_{11}^{y}$	$DP_{11}^{\phi} + DP_{01}^{\phi}$
3	$\overline{LP}_{01}^{x_3}$	$\{\Psi_{01} + \Psi_{11}^{e}\}\hat{\mathbf{x}}$	$LP_{01} + LP_{11}^{ex}$	$HE_{11} + HE_{21}^{e} + TM_{01}$	$TP_{01}^{r} + TP_{11}^{er}$
	$\overline{LP}_{01}^{y_3}$	$\{\Psi_{01} + \Psi_{11}^{e}\}\hat{\mathbf{y}}$	$LP_{01} + LP_{11}^{ey}$	$HE_{11} + HE_{21}^{o} + TE_{01}$	$TP_{01}^{\phi} + TP_{11}^{e\phi}$
4	$\overline{LP}_{01}^{x_4}$	$\{\Psi_{01} + 2\Psi_{11}^{e} + \Psi_{21}^{e}\}\hat{\mathbf{x}}$	$LP_{01} + 2LP_{11}^{ex}$ $+ LP_{21}^{ex}$	$HE_{11} + HE_{21}^{e} + TM_{01}$ $+ LP_{21}^{ex}$	$QP_{01}^{r} + 2QP_{11}^{er}$ $+ QP_{21}^{er}$
	$\overline{LP}_{01}^{y_4}$	$\{\Psi_{01} + 2\Psi_{11}^{e} + \Psi_{21}^{e}\}\hat{\mathbf{y}}$	$LP_{01} + 2LP_{11}^{ey}$ $+ LP_{21}^{ey}$	$HE_{11} + HE_{21}^{o} + TE_{01}$ $+ LP_{21}^{ey}$	$QP_{01}^{\phi} + 2QP_{11}^{e\phi}$ $+ QP_{21}^{e\phi}$

See Figs. 6.1, 6.3, 6.6, and 6.7.

will only be between two supermodes (LP or TP for the respective cases). If we consider excitation with an arbitrary polarization corresponding to excitation of both $\overline{\mathrm{LP}}_{01}^{x_1}$ and $\overline{\mathrm{LP}}_{01}^{y_1}$, then we note that both polarizations of the HE_{11} and HE_{21}-like modes are degenerate. Thus, although six different supermode will be excited, there will only be four rather than six different propagation constants, and thus there will only be six characteristic beat frequencies corresponding to the resulting six different supermode pairs. §

CHAPTER 7

Conclusions and Extensions

In this chapter we first summarize some key results resulting from azimuthal symmetries and longitudinal invariance; then in Secs. 7.2 and 7.3 we discuss extensions to periodic and nonlinear waveguides. Finally, in Secs. 7.4 through 7.8, we discuss some recent developments and periodic guide applications.

7.1 SUMMARY

A major theme of Chaps. 3 to 6 was an examination of what symmetry tells us about the transverse structure of modes of few-mode fibers, using group theory as the mathematical apparatus. In simple cases the consequences of symmetry may be exploited without the mathematical formalism. For example, Snyder and Young [48] used symmetry arguments to construct the true weak-guidance vector modes. Nevertheless, given a group theoretic framework such constructions become trivial and the exploitation of symmetry in more complex cases becomes extremely efficient, especially when the consequences of the symmetries have already been tabulated.

As well as providing an essential building block for more complex structures, this approach provides an interesting tutorial viewpoint for single-core fibers. For example, for a circular core fiber we saw that symmetry dictates that for azimuthal mode number $l = 0$ there is one β level, for $l = 1$ there are three levels, and for $l > 1$ there are two levels.

The group theoretic apparatus used is extremely simple. We have merely used branching rules (often tabulated) to describe the level splitting resulting from symmetry reduction. In particular, reduction of $\mathbf{C}_{\infty v} \otimes \mathbf{C}_{\infty v}$ for LP modes to $\mathbf{C}_{\infty v}$ for true modes gives the above-mentioned polarization splittings. Further reduction to $\mathbf{C}_{2v}$ describes the splitting that occurs when the circular fiber is squashed to an ellipse.

This latter splitting is analogous to the placement of the fiber in a "lattice" of fibers so that it can be regarded as the central fiber. For example, if a fiber is placed in the center of a ring of n symmetrically distributed fibers, then the symmetry of the refractive index "potential" is reduced from $\mathbf{C}_{\infty v}$ to $\mathbf{C}_{nv}$, and the level splitting of the original modes of the central fiber is described by the corresponding branching rule. However, if we have a symmetric array, to construct all the modes we must consider a different method (analogous to, e.g., normal mode analysis in solid-state physics), as explored in Chap. 6.

Knowing the form of the scalar and vector supermode fields determined in Secs. 6.1 to 6.3 enables the evaluation of the splitting of the scalar and vector supermode propagation constants as determined for fundamental mode combinations in Sec. 6.4. Higher-order mode combinations lead to more lengthy calculations and apart from the two-fiber building block were not systematically analyzed here.

Here we have concentrated on azimuthal symmetries described by point groups. If a waveguide is periodic and/or nonlinear, group theory can provide a particularly powerful tool for analysis. Details are beyond the scope of this text; we simply outline a few results.

7.2 PERIODIC WAVEGUIDES

Periodic guides are most commonly analyzed in terms of coupled straight guide modes. Propagation may also be considered in terms of a superposition of *supermodes* or *Bloch modes* of the periodic guide.

To obtain a feel for how longitudinal effects may be included, consider first the case of a guide with *longitudinal invariance*: This longitudinal symmetry is trivially described by the group $\mathbf{T}_1$ of the continuous one-dimensional translations [8 (Chap. 12)]. Group $\mathbf{T}_1$ consists of the identity and $t_{\delta z}$ = operation of translation through an arbitrary distance δz which has one-dimensional matrix representation $D^{(\beta)}(t_{\delta z}) = e^{-i\beta\delta z}$ with basis functions $e^{i\beta z}$, that is, the longitudinal dependence of the field assumed from Chap. 2.

If the guide now has a z dependence with periodicity length L, then only discrete translations remain symmetry operations and it can be shown that the appropriate basis functions are exp $\{i[(\bar{\beta}+n2\pi/L)]z\}$ with n integrals, which gives the z dependence in the Floquet form [66, 68, 125–128] of the periodic field

$$E(\mathbf{r}, z)=\sum_{n=-\infty}^{\infty} a_n\psi_n(\mathbf{r})\exp\left\{i\left(\bar{\beta}\pm n\frac{2\pi}{L}\right)z\right\}$$

Results can be particularly interesting when the z-dependent symmetry is combined with symmetry in the transverse plane for single cores or multicores and the vector nature of the field is considered. This involves taking direct products of the appropriate point and translation groups. (Extensive work has been done in crystallography on resulting space groups for symmetries corresponding to those in a lattice.) One may again consider mode forms under the influence of competing perturbations, using the appropriate symmetry reduction. Techniques are obviously applicable for vector electromagnetic fields in full three-dimensional optical lattices [129, 130].

If the guide is twisted or follows a helicoidal path with pitch L, then the symmetry operation is the screw operation, and it is found that appropriate basis functions are $e^{\pm i(l-n)\phi}e^{[\beta\pm n(2\pi/L)]z}$. All modes of the scalar wave equation are nondegenerate. When vector modes are considered, splitting of the two polarizations of the fundamental mode occurs and is known as Berry's phase (e.g., Refs. 23, 131, and 132 and references therein).

7.3 SYMMETRY ANALYSIS OF NONLINEAR WAVEGUIDES AND SELF-GUIDED WAVES

If the refractive index depends on the field, the wave equation may still possess symmetries. Although the resulting problem is no longer an eigenvalue one, explicit solutions can be obtained by the procedure known as *symmetry reduction of differential equations* [53]. This method is based on the determination of appropriate "symmetry variables" in order to reduce the partial to an ordinary differential equation which is usually simpler to solve. These variables can be of various types, depending on the symmetries present in the equation. In practice, one restricts the consideration to geometric symmetries. For cylindrical (or spherical) symmetry, the symmetry variable will be obviously the radius. For longitudinal symmetry, it will be simply a cartesian coordinate. It could also be an angular coordinate if a dilational symmetry is involved.

Analytic solutions often help provide an understanding of the parameter dependence in addition to being a valuable complement to purely numerical solutions, even if they are for ideal or limiting forms of the index. However, the introduction by the nonlinearity of a longitudinal field-dependent variation in the index leads to multiple types of waves and problems that would be rather intractable without considerable approximation or the above-mentioned symmetry exploitation. As an example application, in the case of the field-dependent index, $n = n_0 + n_1|\mathbf{E}|^2 + n_2|\mathbf{E}|^4$, that is, the "cubic-quintic" nonlinearity [133, 134], the method allowed an extensive classification of the many different types ($\approx$ 100) of exact solutions for the scalar wave equation in the slowly varying amplitude limit, corresponding to cylindrical, longitudinal, dilational symmetries, etc. Nonlinear Schrödinger problems for generalized field-dependent refractive indices [135] are often solved for the many different types of waves in one dimension. The method can also be useful in the analysis of extended models that take into account transverse effects [136], nonlinear dispersion [137], and dissipation or radiation mode [138] effects. Further details are beyond the scope of this chapter, and the reader should consult Ref. 54 and the references therein.

7.4 DEVELOPMENTS IN THE 1990s AND EARLY TWENTY-FIRST CENTURY

The remainder of the chapter briefly discusses some topics that have undergone substantial development since the original manuscript was completed in May 1992.

While optical fibers had been considerably refined since Kao and Hockham's 1966 paper [148] and some may have been tempted by the late 1980s to think that the majority of problems in fiber optics had been solved, the 1990s saw new ideas and an explosion of developments in fiber optics and photonics fueled in part by (1) the telecom boom of the late 1990s and (2) the steady growth of the optical fiber sensor industry [149–152] including novel applications in optofluidics [153–154].

Some of these developments include:

- Ever-more-powerful commercially available modeling software—see Sec. 7.5

- New specialty fiber [215] types (new structures and different materials)
  - Microstructured fibers: holey fibers, photonic crystal fibers (solid-core and air-core PCFs)—these are the subject of Sec. 7.7
  - Practical polymer optical fibers (POFs) [155]: mostly multimode, but some single-mode, and including polymer PCFs and polymer fiber Bragg gratings
  - New rare-earth doped fiber lasers and amplifiers [156, 217, 218]
  - Chiral optical fibers [157]
  - Semiconductor optical fibers such as silicon optical fibers [158] announced in 2008 following on from the development over the last decade of silicon photonics [162]
- New components [160]
  - Particularly those based on fiber Bragg gratings (FBGs) [162]—see Sec. 7.8
  - Nanoprobes such as metal-coated taper aperture probes [16 (Chap. 7)]
- Systematic measurement techniques [162]

Also accompanying the rapid growth in photonics has been an increasing number of good books for students, the R&D community, and users of the technology. Their coverage ranges from electromagnetic waves [164] and photonics overall [165] to fiber optics [166–170], fiber optic components [160, 171], optical waveguides overall [172, 173, 183], photonic crystals [142, 174], photonic crystal fibers [175, 176], anisotropy and nonlinear waveguide phenomena [177], and specialized but important subjects such as polarization mode dispersion [178].

7.5 PHOTONIC COMPUTER-AIDED DESIGN (CAD) SOFTWARE

Since 1992, a range of commercial software packages has become available. These include mode solvers, which can be used alone for longitudinally invariant guides, and propagation packages (that enable modeling of longitudinally varying structures) based on:

- Eigenmode expansion (EME) approaches, particularly bi-directional eigenmode propagation (BEP) [145]
- Beam propagation method (BPM) approaches [179–183]
- Finite difference time domain (FDTD) methods [184]

A full understanding of modes, and, in the case of EME, the modal basis sets used to construct propagating fields, enables much better usage of the modeling tools as well as insight into the physical phenomena they may reveal when used properly. Furthermore, when a modal field needs to be computed many times for longitudinally varying guides, such as for tapered waveguides [185], then exploitation of symmetry to simplify the computation can be particularly advantageous [186, 187]. Full vector approaches such as EME based on vector modes are required for high index contrast guides found in silicon photonics [159]. EME is also particularly suited to optimization of waveguide structures such as tapers [185].

7.6 PHOTONIC CRYSTALS AND QUASI CRYSTALS

Since the original manuscript was completed, there has been substantial development of periodic- and quasi-periodic-guided wave structures [142]:

- A one-dimensional photonic crystal example is a uniform FBG
- Bragg fibers and ARROW fibers:
  - In the 1990s, practical multimaterial Bragg fibers were realized by Fink and coworkers at MIT and Omniguide [216]
  - This work realizes and expands on an idea originally proposed by Yeh, Yariv, and Marom [188] in 1978
  - ARROW (anti-resonance reflecting optical waveguide) fibers [189] and waveguides [190, 191] can have a lower index (air or fluid) core and may find application in optofluidics [153, 154, 189]
  - Rings of varying thickness and oscillating index to achieve a graded effective refractive index were proposed in the late 1970s in a patent by Kao [192]; see also Chap. 5 of this book
- Photonic crystal, microstructured, or holey optical fibers—see Sec. 7.7

Group theory is quite regularly used for the classification of modes for a range of photonic crystal structures [193–199].

7.7 MICROSTRUCTURED, PHOTONIC CRYSTAL, OR HOLEY OPTICAL FIBERS

The invention of microstructured optical fiber, particularly holey fibers in the 1990s [198], has led to a rich literature of analysis, experiments, and applications [175, 176, 200, 201] including, for example:

- "Endless" single-mode operation over a very large wavelength range
- Large mode area operation for high power delivery and fiber laser and amplifier applications
- Ultra low-loss air-core guidance in hollow-core photonic crystal fibers
- Super continuum generation
- Gas-based nonlinear optics

Figure 7.1 shows an example structure, and Fig. 7.2 some modeling for modes of a solid-core photonic crystal fiber [201].

Group theoretic analyses of PCFs have been used in the study of vortex solitons [202] and general numerical analysis [203]. With regard to their azimuthal symmetries, these structures are mostly dealt with in Chaps. 4 and 5 of this book. For example, a key test of the numerical performance of the modeling software for a PCF of hexagonal (triangular) symmetry is the degree to which the two HE_{11} modes are degenerate (as well as the degree of degeneracy for the two HE_{21} modes) [203]. These degeneracies are predicted in Fig. 4.8.

FIGURE 7.1

Example solid-core photonic crystal fiber.

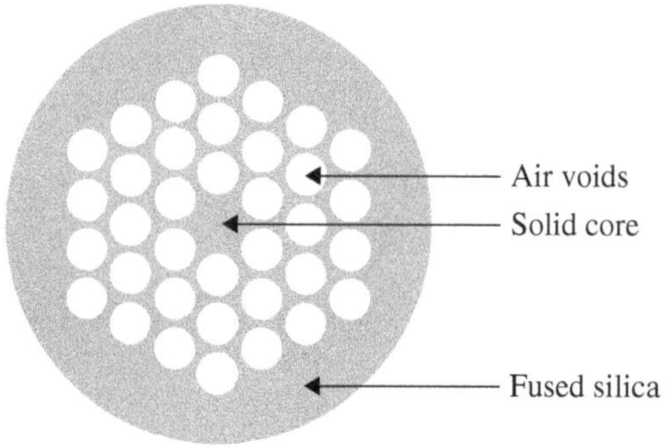

FIGURE 7.2

(*a*) Solid-core photonic crystal fiber model; (*b*) modal field intensity for one of the two degenerate fundamental (HE_{11}-like) modes; (*c*) modal field intensity for one of the two degenerate HE_{21}-like modes which together with the TE_{01}- and TM_{01}-like modes form the second mode set as in Fig. 4.8 [201].

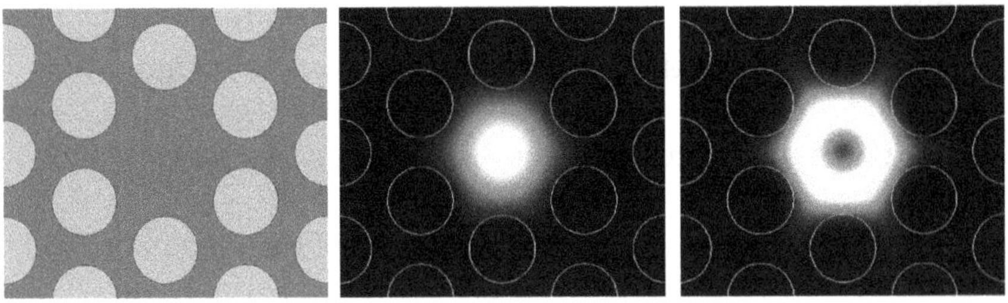

(a) (b) (c)

7.8 FIBER BRAGG GRATINGS

Uniform FBGs (with constant amplitude for the index modulation along their length) are one-dimensional crystals.

While gratings have been analyzed for many years, it was the discovery of laser inscription methods by Hill (initially accidental self-writing [204] and then side-writing using phase masks [205]), Meltz (laser interferometer side-writing [206]) and coworkers [163, 206], Southampton [207, 208], and many other groups that led to the development of FBGs [209, 210, 212] for telecommunications and sensing applications. In the following we consider the conditions for mode reflection and conversion.

7.8.1 General FBGs for Fiber Mode Conversion

Consider two fiber modes with propagation constants β_1 and β_2 related to the modal effective indices $n_{\text{eff}}(\lambda)$ by $\beta = 2\pi\, n_{\text{eff}}(\lambda)/\lambda$. The intermodal beatlength between these two modes is given by

$$z_b \equiv \frac{2\pi}{\beta_1 - \beta_2} = \frac{\lambda}{n_{\text{eff}1} - n_{\text{eff}2}} \tag{7.1}$$

Consider also a grating with periodic refractive index variation and $\Lambda \equiv$ *spatial period* (also referred to as "grating period" [211], "grating spacing" [208], "grating pitch" [163]). For *resonant grating-assisted coupling of the two modes,* we require the grating period to be an integer multiple m_b of the intermodal beatlength z_b for these two modes.

$$\Lambda = m_b z_b \equiv \frac{2\pi m_b}{\beta_1 - \beta_2} = \frac{m_b \lambda}{n_{\text{eff}1} - n_{\text{eff}2}} \tag{7.2}$$

This is general if we regard backward propagating modes as having negative propagation constants. We are usually interested in the first-order ($m_b = 1$) Bragg resonances:

$$\Lambda = z_b \equiv \frac{2\pi}{\beta_1 - \beta_2} = \frac{\lambda}{n_{\text{eff}1} - n_{\text{eff}2}} \tag{7.3}$$

7.8.2 (Short-Period) Reflection Gratings for Single-Mode Fibers

The most commonly used optical fiber gratings are single-mode fiber (SMF) reflection gratings. A schematic is given in Fig. 7.3.

In SMFs, *reflection gratings* are obtained with a grating spacing corresponding to the beatlength between the forward-propagating fundamental mode (β_1) and the backward-propagating fundamental mode ($\beta_2 = -\beta_1$). This Bragg grating spacing is given by

$$\Lambda = z_b = \pi/\beta_1 = \lambda/2\, n_{\text{eff}}(\lambda) \tag{7.4}$$

(This, of course, also applies to reflection of any mode into its backward-propagating form in a multimode fiber.)

If one sends broadband light along a SMF to such a grating, there will be a dip in the transmission spectrum (and a peak in the reflection spectrum) at Bragg wavelength $\lambda_B = 2n_{\text{eff}}\Lambda$.

FIGURE 7.3

Fiber Bragg grating (FBG) for *reflecting* forward-propagating LP_{01} into backward-propagating LP_{01} at Bragg resonance wavelength λ_B.

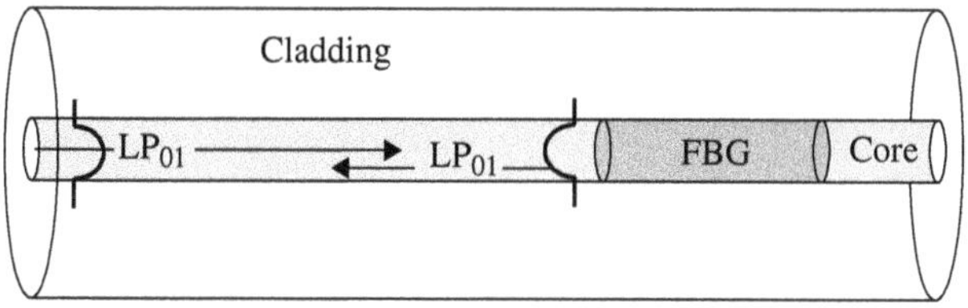

7.8.3 (Long-Period) Mode Conversion Transmission Gratings

If $\beta_1 - |\beta_2| = \delta\beta_{12}$ = difference between propagation constants of two forward-propagating modes, then the Bragg period for conversion between them is given by

$$\Lambda = z_b = 2\pi/\delta\beta_{12} \tag{7.5}$$

In terms of the parameters of Table 2.1, it may be shown that

$$\delta\beta_{12} \approx \frac{2\pi n_{co}\Delta}{\lambda}\frac{U_2^2 - U_1^2}{V^2} = \frac{\lambda(U_2^2 - U_1^2)}{4\pi\rho^2 n_{co}} \text{ and thus } \Lambda \approx \frac{2\rho^2 n_{co}}{\lambda(U_2^2 - U_1^2)} \tag{7.6}$$

For a step index circular fiber, in terms of the core diameter $\phi_{co} = 2\rho$, LP_{11} cutoff wavelength λ_{c11} occurring at $V_{c11} \approx 2.405$,

$$\delta\beta_{12} \approx 0.588\frac{\lambda\sqrt{\Delta}}{\lambda_{c11}\phi_{co}}(U_2^2 - U_1^2) \tag{7.7}$$

Figure 7.7 shows values of U in ranges of interest for optical fibers operated in the few-mode regime (e.g., fibers that are single-mode at telecom wavelengths are few-mode at shorter wavelengths such as in the visible region of the spectrum). Figure 7.8 shows the normalized grating period required to obtain conversion between the fundamental and first lower-order modes. As will be seen in the following example, grating periods for (low-order mode) mode-converting transmission gratings are typically several hundred wavelengths rather than on the order of a third of a wavelength for reflection gratings. Hence the terminology "long-period" and "short-period" gratings for mode-converting transmission and reflecting gratings, respectively.

7.8.4 Example: $LP_{01}\leftrightarrow LP_{11}$ Mode-Converting Transmission FBGs for Two-Mode Fibers (TMFs)

Symmetry consideration: The Bragg planes of the grating need to be tilted with respect to the fiber axis to convert the symmetric mode LP_{01} into the antisymmetric mode LP_{11}. As shown in the schematic of Fig. 7.4, in general both LP_{01} and LP_{11} will exit such a grating.

Demonstration: $LP_{01}\leftrightarrow LP_{11}$ mode-converting transmission FBGs were demonstrated in circular fibers by the Communications Research Center (CRC), Ottawa [213].

FIGURE 7.4

Fiber Bragg grating (FBG) for *converting* forward-propagating LP_{01} into forward-propagating LP_{11} at Bragg resonance wavelength $\lambda_{B01\leftrightarrow11}$. To a first approximation, there will be a sinusoidal power exchange LP_{01} and LP_{11} along the grating.

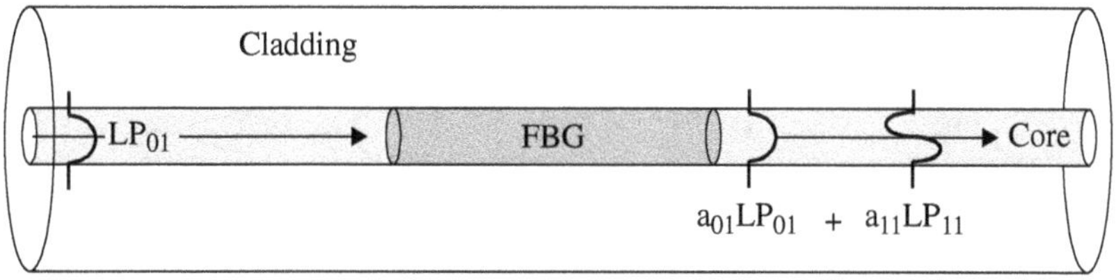

Figure 7.5 shows the required grating period to achieve $LP_{01}\leftrightarrow LP_{11}$ mode conversion for a given wavelength for an optical fiber with parameters similar to that of Ref. 213. Note that for this case the grating periods in the range of 590 to 622 nm lead to two possible wavelengths at which this mode conversion can occur. Thus, for broadband excitation of LP_{01} before the grating, if LP_{11} is stripped off after the grating, two dips will be seen in the wavelength spectrum.

FIGURE 7.5

Phase matching condition for forward-propagating $LP_{01}\leftrightarrow LP_{11}$ conversion for a fiber with parameters similar to that used by Hill [213].

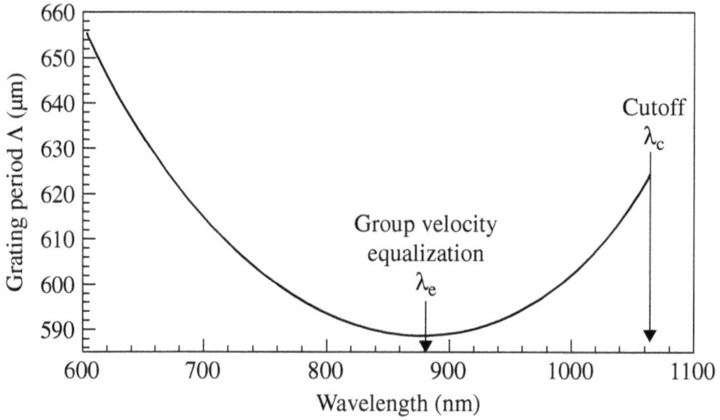

FIGURE 7.6

Fiber Bragg grating (FBG) for *converting* forward-propagating LP_{01} into forward-propagating LP_{01} at Bragg resonance wavelength $\lambda_{01\leftrightarrow 02}$. To a first approximation, there will be a sinusoidal power exchange LP_{01} and LP_{02} along the grating.

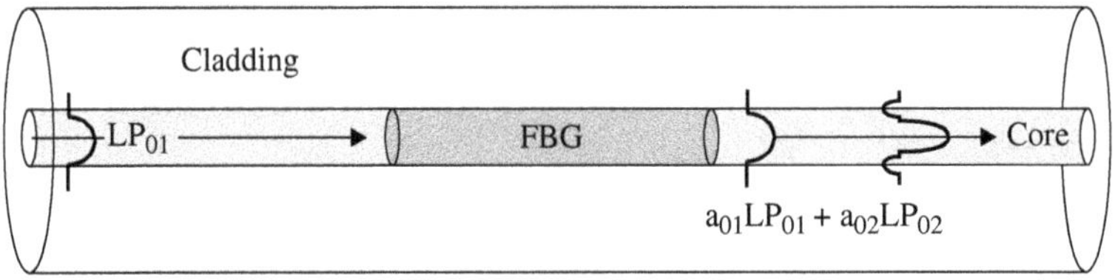

7.8.5 Example: $LP_{01}\leftrightarrow LP_{02}$ Mode-Converting Transmission FBGs

Symmetry consideration: Since LP_{01} and LP_{02} are both circularly symmetric, maximum $LP_{01}\leftrightarrow LP_{02}$ conversion occurs for a nontilted grating. If the grating is tilted then, as well as $LP_{01}\leftrightarrow LP_{02}$ conversion, $LP_{01}\leftrightarrow LP_{11}$ and $LP_{11}\leftrightarrow LP_{02}$ conversions are also possible.

FIGURE 7.7

Normalized modal eigenvalues for circular step-index fibers within the weak guidance approximation obtained by requiring continuity of the scalar field and its first derivative at the core-cladding interface (see also Fig. 1.6).

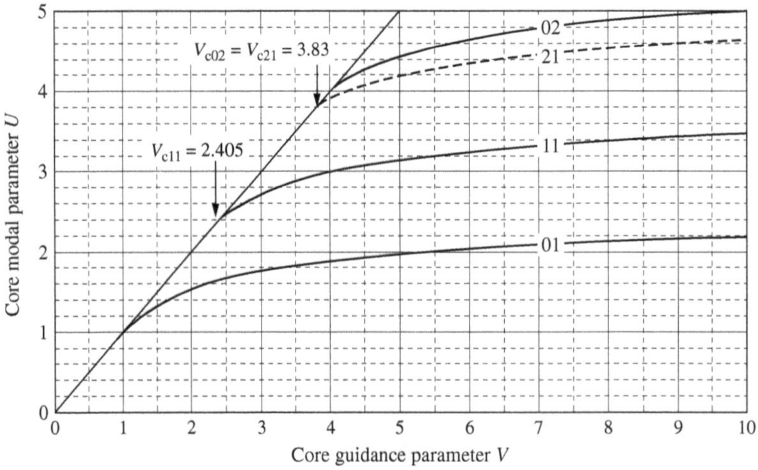

FIGURE 7.8

Grating period required for Bragg conversions between different pairs of the lowest-order modes at given V or λ.

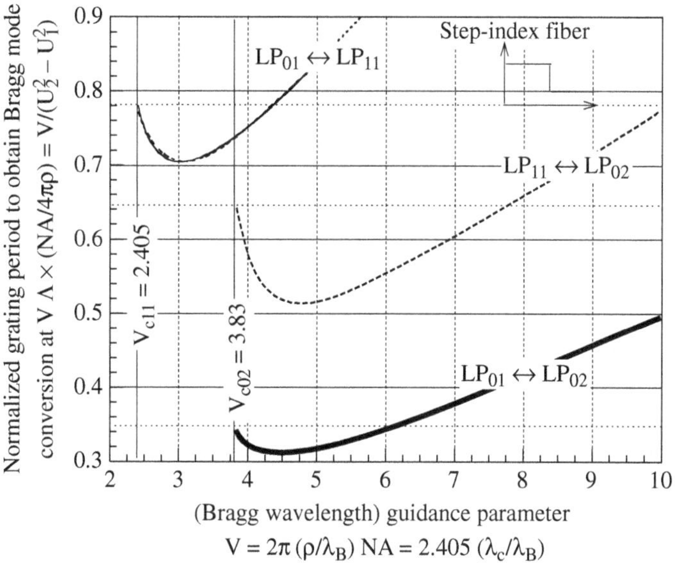

Demonstration in circular fibers: Again the CRC provided early demonstrations of these conversions [214].

Note that in the large V limit we have $U_{0m} = U_{0m}^{\infty} \exp(-1/V)$, where $U_{01}^{\infty} = 2.405$ and $U_{01}^{\infty} = 5.52$, and thus $U_{02}^2 - U_{01}^2 \approx 24.67 \times \exp(-2/V) \to 24.67$.

Note again that, for each Bragg conversion, there is a given pitch range resulting in the conversion occurring at two V-values and thus two wavelengths. For pitch Λ in a range corresponding to $3.8 < V < 6$, the wavelength spectrum for the power remaining in LP_{01} will contain two dips corresponding to $LP_{01} \leftrightarrow LP_{02}$ conversion.

APPENDIX

Group Representation Theory

In this appendix we briefly summarize (1) notation and (2) essential features of group representation manipulations as applied to particular examples in this book. Where the more general formalism is not obvious, the reader is referred to any standard group representation theory text.

§ We should caution particularly the uninitiated student that the hurdles with commencing group theoretic applications are (1) considerable diversity of notation in the literature due to separate development and use for so many different applications and (2) a series of fundamental and essential concepts. Thus, although this appendix should allow reproduction of all results here, our outline is superficial and the caveat remains: for a full understanding, there is no real substitute to the logical progression of working through a text with preliminary definitions and some introductory examples.

There is now an extensive and good selection of well-developed literature, and sufficient familiarity for our purposes may be obtained quite rapidly. For example, for a quick overview, applied mathematics texts such as that by Mathews and Walker [91] include useful summary chapters; Nussbaum [5] has written a particularly readable tutorial paper; Leech and Newman [4] have produced a short text providing a simple introduction to methods of particular relevance, and it is easily followed by undergraduates, as is that by Cracknell [139]. For point group tables ($\mathbf{C}_{nv}$ characters and Kronecker products) and a summary of associated formulas, Atkins et al. [10] have produced a handy 32-page booklet for chemistry and physics students—see Table A.5 for notation. These cover the elementary group representation theory used in the vast majority of this chapter, i.e., for determination of qualitative level splittings and field forms in terms of irrep basis functions. Only symmetry exploitation in quantification of level splitting requires more advanced material involving introduction of the powerful Wigner-Eckart theorem.

These latter topics are covered in volume 1 by Elliot and Dawber [6] which, although it goes considerably further, is also accessible at an undergraduate level; this provides one of the most appropriate references for a thorough understanding of concepts used here including the direct product symmetry reduction methods we use for the inclusion of polarization.

Among the wealth of other more detailed literature, we make particular reference to work by Cornwell [7] that constitutes a highly recommended comprehensive modern text, a classic text by Hamermesh [8], a text by Butler [9] that provides extensive tabulations using a modern notation, and the quantum chemistry text by Cotton [55] that provides extensive applications involving symmetry adapted linear combinations appropriate for multicore fiber problems. For the multicore-core problem, Nussbaum's tutorial [5] is particularly helpful; solid-state applications texts are also relevant, e.g., that by Lax [57] which also includes, e.g., level splitting due to competing perturbations. The text by Stedman [56] includes, e.g., applications to polarization dependence of nonlinear quantum optical processes and references thereto. §

A.1 PRELIMINARIES: NOTATION, GROUPS, AND MATRIX REPRESENTATIONS OF THEM

Although group elements need only be abstract, as is the case for the groups of Table A.1, many groups are defined in terms of sets of entities that happen to satisfy the group multiplication rules, such as

1. *Geometrical symmetry operations* or *coordinate transformations* (e.g., the groups $\mathbf{C}_{nv}$ correspond to rotation and reflection operations)
2. *Matrices*, that is, $g = \mathbb{M}(g)$; for example, $\mathbf{O}_2$, $\mathbf{U}_2$, $\mathbf{SO}_2$, and $\mathbf{SU}_2$ are groups of 2×2 **O**rthogonal or **U**nitary matrices with the **S** for special denoting determinant +1 [7]

TABLE A.1

Group Nomenclature

Group G	Alternative Nomenclature	Description
$\mathbf{C}_{\infty v}$	O(2), $\mathbf{O}_2$	(Continuous) Two-dimensional rotation-reflection group
$\mathbf{C}_{\infty}$	SO(2), $\mathbf{SO}_2$, $\mathbf{O}^+(2)$, $\mathbf{R}_2$[6]	(Continuous) Two-dimensional rotation group
$\mathbf{C}_{nv}$		(Discrete) *n*-Fold rotation-reflection group

3. *Permutations* (for example $\mathbf{S}_n$ is not to be confused with the geometric inversion groups $\mathbf{S}_n$)
4. *Operators*, that is, $g = O(g)$ as here for the specially defined groups $\mathbf{C}^i_{nv}$ being sets of operators corresponding to the group element $\mathbf{C}_{nv}$

Some of these groups satisfy the same multiplication rules and are thus **isomorphic**. For example:

1. $\mathbf{C}_{nv}$ and the permutation groups $\mathbf{S}_n$ for the cases $n = 3$ and 4, and indeed Mathews and Walker [91], analyze triangular problems in terms of the permutations of $\mathbf{S}_3$ rather than the rotation-reflections of $\mathbf{C}_{3v}$ as here.
2. Strictly speaking $\mathbf{O}_2$ is the group of two-dimensional real orthogonal matrices, but as these are just the matrices required to describe rotations and reflections, $\mathbf{O}_2$ is usually regarded as another name for the isomorphic two-dimensional continuous rotation-reflection group $\mathbf{C}_{\infty v}$.
3. The dihedral group $\mathbf{D}_\infty$ is also isomorphic to $\mathbf{C}_{\infty v}$ (and $\mathbf{O}_2$); however, although the same multiplication table applies, the physical operations are quite distinct from those of the two-dimensional rotation-reflection group.

A.1.1 Induced Transformations on Scalar Functions

The transformations $O(R)\psi$ induced on a scalar function of position $\psi(\mathbf{r})$ by a set of coordinate transformations $R \in$ {rotations, reflections} acting on the two-dimensional space $\mathbf{r} = (x, y)$ are defined by

$$O(R)\psi(\mathbf{r}) = \psi(R^{-1}\mathbf{r}) \qquad \text{(A.1)}$$

i.e., transformation of a scalar function is equivalent to an inverse transformation of the coordinates. (Transformation of vector functions is discussed in Sec. A.5.)

A.1.2 Eigenvalue Problems: Invariance and Degeneracies

Consider a system satisfying

$$\mathcal{H}\psi = E\psi \tag{A.2a}$$

where $\mathcal{H}$ is an operator and ψ are eigenfunctions corresponding to eigenvalues E, for example Eq. (2.15) with $\mathcal{H} = \mathcal{H}_s$ and $E = \tilde{\beta}^2$ or Eq. (2.18) with $\mathcal{H} = \mathcal{H}_0$ and $E = \tilde{U}^2$. Here the primary applications of group theoretical methods occur when $\mathcal{H}$ is invariant under a set of symmetry operations $O(g)$ (e.g., rotations, reflections) corresponding to group elements g. In particular, if $O(g)$ commutes with $\mathcal{H}$ [6]

$$O(g)\mathcal{H}\psi_n = \mathcal{H}\,O(g)\psi_n = E_n\,O(g)\,\psi_n \tag{A.2b}$$

then ψ_n and $O(g)\psi_n$ are eigenfunctions corresponding to the same, and thus degenerate eigenvalue E_n; further, group theory provides a systematic tool for exploitation of this symmetry in efficient determination of ψ and the degeneracies of E. Note that for a set of coordinate transformations, invariance of $\mathcal{H} = \mathcal{H}(\mathbf{r})$ means that [6, 8]

$$\mathcal{H}(R\mathbf{r}) = \mathcal{H}(\mathbf{r}) \qquad \text{that is, } O(R)\,\mathcal{H}(\mathbf{r})O(R^{-1}) = \mathcal{H}(\mathbf{r}) \tag{A.2c}$$

§ Wigner-Eckart Theorem for Quantification of Level Splittings

As well as the qualitative information regarding the degeneracies of E trivially obtained once one knows the symmetry group under which $\mathcal{H}$ is invariant, symmetry may lead to quantitative information regarding the magnitudes of level splittings (e.g, due to a perturbation such as ellipticity or inclusion of polarization, see Sec. 3.4.1). Formally this may be obtained by using the Wigner-Eckart theorem. However, application is more complex. We refer to Refs. 6, 7, and 9 for details, remarking that H need not be invariant under the group symmetry operations but just a tensor operator of the group. §

A.1.3 Group Representations

Given a particular group defined in terms of, say, geometrical symmetry operations, one may have *representations* of the group defined in terms of matrices, permutations, etc., *which also obey the same defining group multiplication rules.*

§ However, group representations can differ from groups in that they are sets with elements that need not necessarily be distinct; i.e., the representation is not necessarily "faithful"—e.g., the identity representation labeled **0** has the number 1 corresponding to each group element. Trivially group multiplication is obeyed, although all members of the representation are the same. §

A.1.4 Matrix Irreducible Matrix Representations

A *matrix representation* of the group **G**, which we label $\mathbf{M} \equiv \mathbf{M(G)}$, is a set of matrices $\{\mathbb{D}^{(M)}(g): g \in \mathbf{G}\}$ where the matrices $\mathbb{D}^{(M)}(g)$ have elements $\mathbb{D}^{(M)}(g)_{ij}$ and obey the same set of multiplication rules as the group elements g; that is, $\mathbb{D}^{(M)}(g_1)\,\mathbb{D}^{(M)}(g_2) = \mathbb{D}^{(M)}(g_1 g_2)$ for all $g_1, g_2 \in \mathbf{G}$.

The representation is *irreducible* (an *irrep*) with respect to a certain group **G** if the matrices (corresponding to all elements of that group) *cannot* be *block diagonalized* or broken down in terms of matrices of smaller dimension representations [5, 7, 57]. [Equations (A.4) and (6.3) give example block diagonal matrices.] However, an irrep of **G** may be reducible with respect to a subgroup $\mathbf{G}_s$ (as will be illustrated in Secs. A.3 and A.4).

A.1.5 Irrep Basis Functions

To each irrep **M** of dimension $|M|$ may be associated an irrep basis that is a set of $|M|$ functions $\{\Psi_i^{(M)}: i = 1,\ldots,|M|\}$ (scalar or vector) such that the action on them of the operator $O(g)$ can take the form

$$\boxed{O(g)\Psi_i^{(M)} = \sum\nolimits_j \Psi_j^{(M)} D^{(M)}(g)_{ji}} \tag{A.3}$$

We refer to the symmetry tutorial of Sec. 3.2.4 for an introductory example.

A.1.6 Notation Conventions

We are using notation similar to that of Butler [9] with sets of objects such as groups, vectors, group representations (= sets of,

TABLE A.2

General Group Representation Theory Notation

	Here	Butler [9]	Cornwell [7]	Elliot and Dawber [6]	Hamermesh [8]
General group	**G**	**G**	G	G	G
General group element	g	g, R, S, etc.	T	G	R, S, etc.
Operator	$O(g)$	O_g	P(T)	T(G)	O_R
(Matrix) Irrep	$\mathbf{M} \equiv \mathbf{M(G)}$, (or $\boldsymbol{l}$, ν, etc.)	$\lambda \equiv \lambda(\mathbf{G})$ etc.	Γ^{M}, $\mathbf{D}^{M}$,	$T^{(M)}$, etc. $D^{(M)}$ for $\mathbf{C}_{\infty v}$	$D^{(M)}$
Irrep matrix	$\mathbb{D}^{(M)}(g)$	$\lambda(g)$	$\Gamma^{M}(T)$	$T^{(M)}$	$D^{(M)}(R)$
Irrep matrix element	$D^{(M)}(g)_{ij}$	$\lambda(g)_{ij}$	$\Gamma^{M}(T)_{ij}$	$T_{ij}^{(M)}$	
Irrep basis functions	$\Psi_i^{(M)}$	$\lambda(g)_{ij}$	$\Gamma^{M}(T)_{ij}$	$T_{ij}^{(M)}$	

say, matrices) identified by bold characters, but with several adaptations such as matrices themselves being distinguished by a bold outline character, such as $\mathbb{D}^{(l)}(g)$.

As notations differ considerably, we give Table A.2 to allow an easy correspondence with some of the standard texts. Note in particular that our general matrix representation label **M** corresponds to λ of Butler and the $D^{(M)}$ of many other authors.

A.2 ROTATION-REFLECTION GROUPS

A.2.1 Symmetry Operations and Group Definitions

Here we are particularly concerned with matrix representations of the continuous (infinite) and discrete (finite) rotation-reflection groups $\mathbf{C}_{\infty v}$ and $\mathbf{C}_{nv}$ which, in terms of the symmetry operations of Table A.2, are defined in Table A.3.

A.2.2 Irreps for $\mathbf{C}_{\infty v}$ and $\mathbf{C}_{nv}$

In Tables A.4 and A.5, the continuous group $\mathbf{C}_{\infty v}$ and the finite groups $\mathbf{C}_{nv}$, we give irreducible representation (irrep) matrices $\mathbb{D}^{(M)}(g)$

TABLE A.3a

Rotation-Reflection Group Symmetry Operations

Symmetry Operation Notation Short Full.	Mnemonic	Description
E		**Identity** operation, i.e., no effect: $Eg = gE = g \quad \forall g$
$C_\theta = C_z(\theta)$	Rot(θ)	**Continuous rotation** of angle θ around the symmetry axis, which we choose to be the z axis
C_n		**Discrete rotation** of angle $2\pi/n$ around the z-axis
$\sigma = \sigma_v(xz)$	Rfl	**Vertical-plane reflection**, i.e., in a symmetry plane "vertical" to the plane of rotation and passing through the (z) axis of rotation. *We choose* $\sigma = \sigma_v(xz)$ = reflection in xz plane $= \sigma_y$ = coordinate transformation mapping (x, y) to $(x, -y)$

TABLE A.3b

Rotation (-Reflection) Group Definitions in Terms of Symmetry

$\mathbf{C}_{\infty\mathbf{v}} = \{EC_\theta, \sigma C_\theta: -\pi \le \theta \le \pi\}$
= infinite group with three classes of symmetry operation: (1) identity E, (2) *continuous* rotation C_θ , (3) vertical-plane reflection σ_v.

$\mathbf{C}_{n\mathbf{v}} = \{EC_n^m, \sigma C_n^m: m = 0, 1,..,n-1\} = \{E, C_n, C_n^2, ..., C_n^{n-1}, \sigma, \sigma C_n, \sigma C_n^2, ..., \sigma C_n^{n-1}, \}$
= finite group with $2n$ elements being powers and products of (1) identity E, (2) *discrete* $2\pi/n$ rotation C_n, (3) vertical-plane reflection σ_v

$\mathbf{C}_\infty = \{EC_\theta : -\pi \le \theta \le \pi\}$; $\mathbf{C}_n = \{EC_n^m: m = 0,1,..,n-1\} = \{E, C_n, C_n^2, , ..., C_n^{n-1}\}$

corresponding to irreps **M** with the irrep labeling notation being described in the next subsection.

Furthermore, for $\mathbf{C}_{\infty\mathbf{v}}$ in Table A.4, we give the irrep matrices for real and complex bases together with corresponding basis functions that satisfy Eq. (A.3). For example, a set of scalar basis functions (SBFs) corresponding to the real irreps $\mathbf{M}(\mathbf{C}_{\infty\mathbf{v}})$ is $\{\Psi_1^{(M)}, \Psi_2^{(M)}\} = \{\cos M\phi, \sin M\phi\}$. A set of vector basis functions (VBFs) corresponding to the irrep $\mathbf{1}(\mathbf{C}_{\infty\mathbf{v}})$ is $\{\hat{\mathbf{x}}, \hat{\mathbf{y}}\}$. In the rest of this appendix we concentrate on results for the real irreps (as required

TABLE A.4

Real and Complex Irreducible Matrix Representations (Irreps) of $C_{\infty v}$

Irrep Label, M		Symmetry operations, g and Irrep Matrices $\mathbb{D}^{(M)}(g)$			ϕ-Dependent Basis Functions All BFs may be multiplied by general Function $F(r)$	
Natural Notation	**Mul. Notation**	E	C_θ	$\sigma = \sigma_v(xz)$	**Scalar BFs** $\Phi_s^M(\phi)$	**Vector BF's**
0	$\mathbf{A_1}$	1	1	1	1	$\hat{\mathbf{r}}$
$\tilde{0}$	$\mathbf{A_2}$	1	1	−1		$\hat{\phi}$
b = real **M** (≥**1**)	$\mathbf{E_M}$	$\begin{bmatrix} 1 & 0 \\ 0 & 1 \end{bmatrix}$	$\begin{bmatrix} \cos M\theta & -\sin M\theta \\ \sin M\theta & \cos M\theta \end{bmatrix}$	$\begin{bmatrix} 1 & 0 \\ 0 & -1 \end{bmatrix}$	$M = 1$: $\{\cos\phi, \sin\phi\} \sim \{x, y\}$ $M \geq 1$: $\{\cos M\phi, \sin M\phi\}$	$\{\hat{\mathbf{x}}, \hat{\mathbf{y}}\}$ $\{\hat{\mathbf{p}}_b^{Ml} : h = 1,2\}_{b=r}^{l=M\pm1}$
b = complex **M** (≥**1**)	$\mathbf{E_M}$	$\begin{bmatrix} 1 & 0 \\ 0 & 1 \end{bmatrix}$	$\begin{bmatrix} e^{+iM\theta} & 0 \\ 0 & e^{-iM\theta} \end{bmatrix}$	$\begin{bmatrix} 0 & 1 \\ -1 & 0 \end{bmatrix}$	$M = 1$: $\{e^{+i\phi}, e^{-i\phi}\}$ $M \geq 1$: $\{e^{+iM\phi}, e^{-iM\phi}\}$	$\{\hat{\mathbf{R}}, \hat{\mathbf{L}}\}$ $\{\hat{\mathbf{p}}_h^{Ml} : h = 1,2\}_{b=c}^{l=M+1}$

Vector basis functions (VBF's)

For $M \geq 1$, for both the real and complex irreps there are two sets of ϕ-dependent VBFs that satisfy Eq. (A.3), that is $\{\hat{\mathbf{p}}_1^{Ml}\hat{\mathbf{p}}_2^{Ml}\}$ with (i) $l = M + 1$ and (ii) $l = M - 1$ where the vector cylindrical harmonics are defined as $\hat{\mathbf{p}}_h^{Ml}(\phi) \equiv c_{hsk}^{Ml}\Phi_s^l(\phi)\hat{\mathbf{p}}_k$ with $\Phi_s^l(\phi)$ the SBFs, and $\hat{\mathbf{p}}_k = \{\hat{\mathbf{x}}, \hat{\mathbf{y}}\}$ for b = real and $\{\hat{\mathbf{R}}, \hat{\mathbf{L}}\}$ for b = complex and c_{hsk}^{Ml} correspond to the CG coefficients of Sec. A4.

TABLE A.5

Real Irreducible Matrix Representations of C_{nv}

Irrep Label M					Symmetry Operations, g			
	B-K Irrep $\Gamma^{(i)}$ label i				E	$\{C_n^m;\, m = 1,\ldots,n-1\}$	$\sigma = \sigma_v(xz)$	$\{\sigma C_n^m;\, m = 1,\ldots,n-1\}$
Nat. (Butler)	**Mul.**	$n = 2$	**3**	**4, 6**	Irrep Matrices, $\mathbb{D}^{(\mathbf{M})}(g)$			
0	A_1	1	1	1	1	1	1	1
$\tilde{0}$	A_2	3	2	2	1	1	–1	–1
$1 \le \mathbf{M} < \frac{n}{2}$	E_M	—	3	$4 + p$	$\begin{bmatrix} 1 & 0 \\ 0 & 1 \end{bmatrix}$	$\begin{bmatrix} \cos M\theta_m & -\sin M\theta_m \\ \sin M\theta_m & \cos M\theta_m \end{bmatrix}$	$\begin{bmatrix} 1 & 0 \\ 0 & -1 \end{bmatrix}$	$\begin{bmatrix} \cos M\theta_m & -\sin M\theta_m \\ -\sin M\theta_m & -\cos M\theta_m \end{bmatrix}$
If n even:								
(n/2)	B_1	2	—	3	1	$(-1)^m$	1	$(-1)^m$ (n even)
$(\widetilde{n/2})$	B_2	4	—	4	1	$(-1)^m$	–1	$-(-1)^m$ (n even)

The angle $\theta_m = 2\pi m/\tilde{n}$, and $M = 1, 2, \ldots, M_n$ with $M_n = n/2$ one for n even or $(n-1)/2$ for n odd.

for the analysis of LP and standard hybrid mode constructions). For more detailed discussion concerning the complex irrep results (used for the analysis of circularly polarized mode forms), when not obvious, we refer to Ref. 1 and/or a standard text such as that by Hamermesh [8].

§ A.2.3 Irrep Notation

1 **Nat.:** Here we use the **natural** general labeling of irreps with M corresponding to the irrep label used in the tables of **Butler** [9].

2. **Mul.: Mullikan's notation: A_n, B_n** for one-dimensional **irreps** (**A/B** denoting principal rotation ±1 and **1/2** denoting reflection ±1) and $\mathbf{E_n}$ for **two-dimensional irreps** is the labeling favored particularly in molecular physics and chemistry and used in Hamermesh [8] and the set of tables [10].

3. **For groups C_{nv}: B-K notation** $\Gamma^{(i)}$ with i listed above for particular values of n (= 2, 3, 4, 6) corresponding to the crystal point groups provides an older labeling (originally due to **Bethe** and used in extensive tabulations of **Koster et al.**) popular in crystallography and solid-state physics [57]. See Table A.1 of Ref. 56 for a more detailed list of corresponding irreps. §

A.3 REDUCIBLE REPRESENTATIONS AND BRANCHING RULE COEFFICIENTS VIA CHARACTERS

The modal *level splitting* in this chapter resulting from the reduction of symmetry from one group to another one is determined by using matrix representation branching rules. In many cases the consequences of such symmetry reductions have been already calculated and are conveniently tabulated. In any case they are trivially calculated. We provide a brief explanation.

As mentioned in Sec. A.1.5, matrix representations are said to be *reducible* if they can be *block-diagonalized* or, more specifically, broken down in terms of matrix representations of smaller dimension. The corresponding *reductions* may be described by *branching rules*. Furthermore, although a representation may be irreducible with respect to a certain group when all the symmetry operations of that group are considered (e.g., the continuous rotation-reflections of $\mathbf{C}_{\infty v}$), if the operations are restricted to those of a subgroup, (e.g., just the π rotation-reflections) of $\mathbf{C}_{2v}$, then the group representations may be reducible with respect to the subgroup representations. We denote such group-subgroup reductions by the subset symbol $\supset$ (for example, $\mathbf{C}_{\infty v} \supset \mathbf{C}_{2v}$ which is sometimes written as the *subduction* $\mathbf{C}_{\infty v} \downarrow \mathbf{C}_{2v}$, as in Ref. 40.

A.3.1 Example Branching Rule for $\mathbf{C}_{\infty v} \supset \mathbf{C}_{2v}$

As a particular example, it may be verified from Tables A.4 and A.5

$$\mathbb{D}^{(2)}_{\infty v}(g) = \begin{bmatrix} \mathbb{D}^{(0)}_{2v}(g) & 0 \\ 0 & \mathbb{D}^{(\tilde{0})}_{2v}(g) \end{bmatrix} \equiv \mathbb{D}^{(0)}_{2v}(g) \oplus \mathbb{D}^{(\tilde{0})}_{2v}(g) \qquad \forall g \in \mathbf{C}_{2v} \quad \text{(A.4}a\text{)}$$

where the **bold subscripts refer to the groups** ($\mathbf{C}_{\infty v}$ or $\mathbf{C}_{2v}$). Thus we see that when we consider only those group elements of $\mathbf{C}_{\infty v}$ also in the subgroup $\mathbf{C}_{2v}$, the corresponding matrices for the irrep $\mathbf{2}(\mathbf{C}_{\infty v})$ are *block-diagonal;* that is **2** is *reducible with respect to* $\mathbf{C}_{2v}$. Equation (A.4*a*) may be written as the **branching rule**

$$\mathbf{2}(\mathbf{C}_{\infty v}) \rightarrow \mathbf{0}(\mathbf{C}_{2v}) \oplus \tilde{\mathbf{0}}(\mathbf{C}_{2v}) \qquad \text{or abbreviated as} \qquad \mathbf{2}_{\infty v} \rightarrow \mathbf{0}_{2v} \oplus \tilde{\mathbf{0}}_{2v} \qquad \text{(A.4}b\text{)}$$

This has application to the splitting of the HE_{2m} modes when a circular fiber is elliptically deformed as discussed in Sec. 4.2.1 and Fig. 4.3.

Mode degeneracy determination for multicore fibers (Sec. 6.2.1.) is another application that involves block-diagonal or reducible matrices. However, in this case the reducible matrix representations are not obtained as irreducible representations of a higher symmetry group.

A.3.2 BRANCHING RULE COEFFICIENTS VIA CHARACTERS

§ The $\mathbf{C}_{\infty v} \supset \mathbf{C}_{2v}$ reduction of two- to one-dimensional matrices is in fact quite special in that, in general, (1) a *diagonalizing* transformation of the form $\mathbb{T}\,\mathbb{D}(g)\,\mathbb{T}^{-1}$ is required and (2) for reducibility, after such a transformation it is only necessary to have zero matrix elements below the matrices on the diagonal (i.e., in the most general case it is possible to have nonzero irrep matrices in the upper right). §

However, the actual branching rules may usually be simply obtained from tables or, alternatively, in terms of branching rule coefficients that are trivially calculated via the appropriate group *characters*.

In particular, consider the branching rules for the symmetry reduction $\mathbf{C}_{\infty v} \supset \mathbf{C}_{nv}$, which we write as

$$l(\mathbf{C}_{\infty v} \to \sum_{M} \oplus n_M^l \mathbf{M}(\mathbf{C}_{nv}) \quad \text{or abbreviated as} \quad \mathbf{l}_{\infty v} \to \sum_{M} \oplus n_M^l \mathbf{M}_{nv} \tag{A.5}$$

The number n_m^l of times the irreducible representations $\mathbf{M(C_{nv})}$ appear in the irreps of $l(\mathbf{C}_{\infty v})$ is given by the general relation [7, 8]

$$n_M^l = \frac{1}{|G|} \sum_{g \in G} \chi^{(l)}(g) \chi^{(M)}(g)^* \tag{A.6}$$

where $|G|$ is the number of elements g (order) of the group (here $\mathbf{C}_{nv}$), the asterisk* signifies complex conjugate (no importance for $\mathbf{C}_{nv}$ or $\mathbf{C}_{\infty v}$), and the group character is defined as

$$\chi^{(M)}(g) = \sum_{j=1}^{|M|} D^{(M)}(g)_{jj} = \text{trace } \mathbb{D}^{(M)}(g) \qquad |M|$$

$$= \text{dimension of matrix irrep } \mathbf{M} \tag{A.7}$$

Characters are tabulated in most texts on group representation theory. Note that the character of the identity element gives the dimension of the (matrix) irrep (e.g., compare the matrices of Tables A.4 and A.5 with characters for *E* in Tables A.6 and A.7). For one-dimensional irreps (e.g., all those of $\mathbf{C}_{2v}$) the matrix representations of the elements are simply the characters.

TABLE A.6

Characters for $C_{\infty v}$

Irrep	E	C_θ	σ
0	1	1	1
$\tilde{\mathbf{0}}$	1	1	−1
M ≥ **1**	2	$2\cos M\theta$	0

TABLE A.7

Character (and Irrep) Table for C_{2v}

Irrep			Group Elements/ Classes				φ-Dependent Basis Functions	
Nat.	*Mull.*	*B-K*	*E*	C_2	σ	σC_2	Scalar(SBFs) m = integer	Vector (VBFs)
0	A_1	$\Gamma^{(1)}$	1	1	1	1	$\cos 2m\phi$	$\hat{r}$
$\tilde{0}$	A_2	$\Gamma^{(3)}$	1	1	−1	−1	$\sin 2m\phi$	$\hat{\phi}$
1	B_1	$\Gamma^{(2)}$	1	−1	1	1	$\cos (2m+1)\phi$	$\hat{x}$
$\tilde{1}$	B_2	$\Gamma^{(4)}$	1	−1	−1	−1	$\sin (2m+1)\phi$	$\hat{y}$

As for all groups C_{nv}, r is invariant under rotations/reflection; thus all basis functions (BFs) can be multiplied by a general function $F(r)$.

Example 1: $C_{\infty v} \supset C_{2v}$

§ For $C_{\infty v} \supset C_{2v}$, given the character tables for the two groups as Tables A.6 and A.7, the branching rules of Table A.8 required for the circular-to-elliptical symmetry reduction of Fig. 4.3 are obtained. In particular, the representations l, which are two-dimensional for $l \geq 1$ are reduced in terms of one-dimensional representations M_{2v}. §

Example 2: Product group reduction $C_{\infty v} \otimes C_{\infty v} \supset C_{\infty v}$

§ To achieve the reduction of the *direct* or *Kronecker* product representations $1 \otimes 1(C_{\infty v} \otimes C_{\infty v})$ into irreducible representations $\nu(C_{\infty v})$, we note that the characters of the product group are simply given

TABLE A.8

Branching Rules for the Symmetry Reduction $C_{\infty v} \supset C_{2v}$

	Characters, $\chi^{(l)}(g)$, of elements g of $C_{\infty v}$ in C_{2v}				Resolution of $(C_{\infty v})$ into irreps of $M(C_{2v})$
Irreps $l(C_{\infty v})$	E	C_2	σ	σC_2	
0	1	1	1	1	0
$\tilde{0}$	1	1	−1	−1	$\tilde{0}$
1	2	−2	0	0	$1 \oplus \tilde{1}$
2	2	2	0	0	$\tilde{0} \oplus \tilde{0}$
3	2	−2	0	0	$\tilde{1} \oplus \tilde{1}$
⋮	⋮	⋮	⋮	⋮	⋮

by the product of the individual group characters of $\mathbf{C}_{\infty v}$ in Table A.6. The coefficient is simply calculated from the relation

$$n_{l1}^{v} = \int_{\mathbf{G}} \chi^{(l)}(g)\chi^{(1)}(g)\chi^{(v)*}(g)dg \tag{A.8}$$

where the invariant integration [7] for $\mathbf{C}_{\infty v}$ is given by

$$\int f(g)dg = \frac{1}{2\pi \mid g_{\text{discrete}} \mid} \sum_{g_{\text{discrete}}} \int_{\theta=\pi}^{\pi} f(C_\theta g_{\text{discrete}})d\theta \qquad g_{\text{discrete}} \in \{E, \sigma_v\} \tag{A.9}$$

For C_{∞}, $g_{\text{discrete}} \in \{E\}$, and for the finite groups $\mathbf{C}_{nv}$ the invariant integration is simply replaced by the discrete sum $1/|G| \; \Sigma_{g \in G}$ as used in Eq. (A.7). Note that Eq. (A.8) simply generalizes Eq. (A.7) to the case where the starting group is a direct product of two groups. Note also that for many product group reductions these coefficients $n_{l1}^{v} = \{l1v\}$, sometimes referred to as "3-*j* phases" [9], are already tabulated. For example, for $\mathbf{C}_{nv} \otimes \mathbf{C}_{nv} \supset \mathbf{C}_{nv}$ with n = (2, 3, 6), 4, 5, and ∞, see the Direct Product Tables 2, 4, 5, and 10, respectively, on pp. 14 to 17 of Atkins et al. [10]. §

A.4 CLEBSCH-GORDAN COEFFICIENT FOR CHANGING BASIS

§ Equation (3.18) expresses the basis functions of the representations ν of the reduced product group $C_{\infty v} \otimes C_{\infty v} \supset C_{\infty v}$ in terms of products of the basis functions of the individual group representations **l** and **1** via the Clebsch-Gordan (CG) coefficients. One method of calculating CG coefficients (if tables are not available) is to use the following relation (valid for unitary [7] representations)

$$\sum_{sk} D^{(l)}(g)_{s's} D^{(1)}(g)_{k'k} < ls, 1k \mid vh >$$

$$= \sum_{h} < ls', 1k' \mid vh' > D^{(v)}(g)_{h'h} \qquad \forall g \in \mathbf{C}_{\infty v} \tag{A.10}$$

where the matrix elements $D^{(l)}(g)_{s's}$ for the appropriate irreps of $\mathbf{C}_{\infty v}$ have been given in Table A.4. The CG coefficients of interest for

TABLE A.9

CG Coefficients $c^{\nu l1}_{hsk} = \langle ls, 1k \, \nu h \rangle$ for Real Representation Reduction $l = \otimes 1 \to \sum_{\nu} \oplus n^{\nu}_{l1}\nu$ of $\mathbf{C}_{\infty v} \otimes \mathbf{C}_{\infty v} \supset \mathbf{C}_{\infty v}$ within Normalization $1/\sqrt{2}$ and Phase Factor

	νh						(l>1)			
ls*, 1*k	**01**	**$\tilde{0}$1**	**11**	**12**	**21**	**22**	***l*−1, 1**	***l*−1, 2**	***l*+1, 1**	***l*+1, 2**
01, 11			1	0						
01, 12			0	1						
11, 11	1	0			1	0				
11, 12	0	1			0	1				
12, 11	0	−1			0	1				
12, 12	1	0			−1	0				
*l*1, 11							1	0	1	0
*l*1, 12							0	1	0	1
*l*2, 11							0	−1	0	1
*l*2, 12							1	0	−1	0

the real representation reduction $l \otimes 1(\mathbf{C}_{\infty v} \otimes \mathbf{C}_{\infty v})$ in terms of $\nu(\mathbf{C}_{\infty v})$ are given in Table A.9.

Strictly speaking, orthonormality conditions (e.g., Ref. 9) require the absolute magnitude of each of these CG coefficients to be $1/\sqrt{2}$. This corresponds to the fact that to have the same power normalization we would require, e.g.,

$$\mathrm{HE}^{e}_{l+1m} = \frac{1}{\sqrt{2}}\left\{\mathrm{LP}^{ex}_{lm} - \mathrm{LP}^{oy}_{lm}\right\}$$

$$= \langle l1, 11 \,|\, l+11 \rangle \mathrm{LP}^{ex}_{lm} + \langle l2, 12 \,|\, l+11 \rangle \mathrm{LP}^{oy}_{lm} \qquad \text{(A.11) §}$$

A.5 VECTOR FIELD TRANSFORMATION

§ Our treatment of polarization has parallels to that used for incorporating spin in quantum mechanics, carefully explained, e.g., in Elliot and Dawber [6].

In comparison with the scalar field transformation of Eq. (A.1), each component of a transverse vector field [i.e., a tensor field of degree 1 in the two-dimensional version of Eq. (16.18) of Ref. 6] transforms under reflection/rotations $R \in \mathbf{C}_{\infty\mathbf{v}}$ as [6, 7]

$$O_J(R)e_j(\mathbf{r}) = \sum_{k=1}^{2} D^{(1)}(R)_{jk} e_k(R^{-1}\mathbf{r}) \tag{A.12a}$$

This is the vector version of the induced (passive) transformation of Eq. (A.1). It follows from application of $O_S(R)$ and $O_P(R)$ to the scalar magnitude and polarization, respectively, i.e.,

$$O_J(R)\mathbf{e}_t(\mathbf{r}) = \sum_{k=1}^{2} O_S(R)\{e_k(\mathbf{r})\} \quad O_P(R)\{\hat{\mathbf{p}}_k\} = \sum_{k=1}^{2} e_k(R^{-1}\mathbf{r})(\hat{\mathbf{p}}_k)' \tag{A.12b}$$

where from Eqs. (A.1) and (A.3)

$$O_S(R)e_k(\mathbf{r}) = e_k(R^{-1}\mathbf{r}) \qquad \text{and} \qquad (\hat{\mathbf{p}}_k)' \equiv O_P(R)\hat{\mathbf{p}}_k = \sum_{k=1}^{2} D^{(1)}(R)_{jk}\hat{\mathbf{p}}_j \tag{A.12c}$$

Thus

$$O_J(R)\mathbf{e}_t(\mathbf{r}) = \sum_{k=1}^{2} e_k(R^{-1}\mathrm{r}) \sum_{j=1}^{2} D(C_\theta)_{jk}\hat{\mathbf{p}}_j = \sum_{j=1}^{2} \hat{\mathbf{p}}_j O_J(R)\{e_j(\mathrm{r})\} \tag{A.12d}$$

with $O_J(R)e_j(\mathbf{r})$ as given above. §

REFERENCES

1. Black, R. J., Gagnon, L., and Stedman, G. E., "Symmetry Principles for Lightguides: I. Single-Core Few-Mode Fibers," unpublished manuscript, 1992.
2. Gagnon, L., and Black, R. J., "Symmetry Principles for Lightguides: II. Fiber Arrays," unpublished manuscript, 1992.
3. Snyder, A. W., and Love, J. D., *Optical Waveguide Theory*, Chapman and Hall, London, 1983.
4. Leech, J. W., and Newman, D. J., *How to Use Groups*, Methuen, London, 1969.
5. Nussbaum, A., "Group Theory and Normal Modes," *Am. J. Phys.* **36**(6), 529 (1968).
6. Elliot, J. P., and Dawber, P. G., *Symmetry in Physics*, vol. 1: *Principles and Simple Applications*, vol. 2: *Further Applications*, MacMillan, London, 1979.
7. Cornwell, L. F., *Group Theory in Physics*, Academic, London, 1984.
8. Hamermesh, M., *Group Theory and Its Applications to Physical Problems*, Addison-Wesley, Reading, Mass., 1964.
9. Butler, P. H., *Point Group Symmetry Applications*, Plenum, New York, 1981.
10. Atkins, P. W., Child, M. S., and Phillips, C. S. G., *Tables for Group Theory*, Oxford University Press, Oxford, U.K., 1970.
11. Snitzer, E., "Cylindrical Dielectric Waveguide Modes," *J. Opt. Soc. Am.* **51**, 491 (1961).
12. Dyott, R. B., "Cut-off of the First Higher Order Modes in Elliptical Dielectric Waveguide: An Experimental Approach," *Electron. Lett.* **26**(20), 1721 (1990).
13. Marcuse, D., *Dielectric Optical Waveguides*, Academic, New York, 1991.
14. Black, R. J., Hénault, A., Gagnon, L., Gonthier, F., and Lacroix, S., "Intensity-Dependent Few-Mode Lightguide Interferometry," *Topical Meeting on Nonlinear Guided-Wave Phenomena: Physics and Applications*, Black, R. J., Hénault, A., Gagnon, L., Gonthier, F., and Lacroix, S. (eds.), *1989 Technical Digest Series*, vol. 2, Optical Society of America, Washington, D.C. (1989), p. 66.
15. Garth, S. J., and Pask, C., "Nonlinear Polarization Behaviour on Circular Core Bimodal Optical Fibers," *Electron. Lett.* **25**(3), 182 (1989).
16. Garth, S. J., and Pask, C., "Polarization Rotation in Nonlinear Bimodal Optical Fibers," *J. Lightwave Technol.* **8**(2), 129 (1990).
17. Garth, S. J., and Pask, C., "Nonlinear Effects in Elliptical-Core Few-Mode Optical Fibers," *J. Opt. Soc. Am. B* **9**(2), 243 (1992).
18. Stolen, R. H., and Bjorkholm, J. E., "Parametric Amplification and Frequency Conversion in Optical Fibers," *IEEE J. Quantum Electron.* **QE-18**, 1062 (1982).
19. Hill, K. O., Malo, B., Vineberg, K. A., Bilodeau, F., Johnson, D. C., and Skinner, I. M., "Efficient Mode Conversion in Telecommunication Fibre Using Externally Written Gratings," *Electron. Lett.* **26**(16), 1270 (1990).
20. Skinner, I. M., "Average Higher-Order Modes of a Periodically, Axi-asymmetrically Perturbed Optical Fibre," unpublished manuscript, 1991.
21. Snyder, A. W., "Physics of Vision in Compound Eyes," in H. Autrum (ed.), *Handbook of Sensory Physiology*, Springer-Verlag, Berlin, 1979, p. 225.

22. Eftimov, T. A., "Resultant Mode Pattern and Polarization in a LP_{01}-LP_{02} Two-Mode Linearly Birefringent Optical Fibre," *Opt. & Quantum Electron.* **23**, 1143 (1991).
23. Lipson, S. G., "Berry's Phase in Optical Interferometry: A Simple Derivation," *Opt. Lett.* **15**(3), 154 (1990).
24. Archambault, J.-L., Black, R. J., Bures, J., Gonthier, F., Lacroix, S., and Saravanos, C., "Few-Mode Interferometry: A New, Simple, and Accurate Characterization Method for Telecommunications Fibers," *OFC'91, Optical Fiber Communications*, San Diego, Calif., *1991 Technical Digest Series* **4** (Optical Society of America, Washington, D.C.), 116 (1991).
25. Archambault, J.-L., Black, R. J., Bures, J., Gonthier, F., Lacroix, S., and Saravanos, C., "Fiber Core Profile Characterization by Measuring Group Velocity Equalization Wavelengths," *IEEE Phot. Technol. Lett.* **3**(4), 351 (1991).
26. Huang, S. Y., Blake, J. M., and Kim, B. Y., "Perturbation Effects on Mode Propagation Effects in Highly Elliptical Core Two-Mode Fibers," *J. Lightwave Technol.* **8**(1), 23 (1990).
27. Murphy, K. A., Miller, M. S., Vengsarkar, A. M., and Claus, R. O., "Elliptical-Core Two-Mode Optical-Fiber Sensor Implementation Methods," *J. Lightwave Technol.* **8**(11), 1688 (1990).
28. Black, R. J., Gonthier, F., Lacroix, S., Lapierre, J., and Bures, J., "Abruptly Tapered Fibers: Index Response for Sensor Application," *4th International Conference on Optical Fiber Sensors, OFS '86*, Tokyo, *Technical Digest, Post-Deadline Papers*, PD-3 (1986).
29. Kim, B. Y., Blake, J. N., Huang, S. Y., and Shaw, H. J., "Use of Highly Elliptical Core Fibers for Two-Mode Fiber Devices," *Opt. Lett.* **12**, 729 (1987).
30. Lacroix, S., Gonthier, F., Black, R. J., and Bures, J., "Interferometric Properties of Tapered Monomode Fibres," *13th European Conference on Optical Communication, ECOC '87*, Helsinki, *Tech. Digest* **I**, 219 (1987).
31. Black, R. J., Gonthier, F., Lacroix, S., Lapierre, J., and Bures, J., "Tapered Fibers: An Overview" (invited paper), *Symposium on Fiber Optics and Optoelectronics, O-E/Fibers '87*, San Diego, V. J. Tekippe (ed.), *Components for Fiber Optic Application II, Proc. SPIE.* **839**, 2 (1987).
32. Black, R. J., Gonthier, F., Lacroix, S., and Love, J. D., "Tapered Single-Mode Fibres and Devices: Part 2. Experimental and Theoretical Quantification," *IEE Proc. Pt. J: Optoelectronics* **138**(5), 355 (1991).
33. Gonthier, F., Lacroix, S., Daxhelet, X., Black, R. J., and Bures, J., "Broadband All-Fiber Filters for Wavelength-Division-Multiplexing Application," *Appl. Phys. Lett.* **54**(14), 1290 (1989).
34. Kumar, A., Das, U. K., Varshney, R. K., and Goyal, I. C., "Design of a Dual-Mode Filter Consisting of Two Dual-Mode Highly Elliptical Core Fibers," *J. Lightwave Technol.* **8**(1), 34 (1990).
35. Hill, K. O., Malo, B., Johnson, D. C., and Bilodeau, F., "A Novel Low-Loss Inline Bimodal-Fiber Tap: Wavelength-Selective Properties," *IEEE Phot. Technol. Lett.* **2**(7), 484 (1990).
36. Blake, J. N., Kim, B. Y., Engan, H. E., and Shaw, H. J., "All-Fiber Acousto-Optic Frequency Shifter Using Two-Mode Fiber," *Fiber Optic Gyro: 10th Anniversary Conference*, **719**, 92 (1986).

37. Penty, R. V., White, I. H., and Travis, A. R. L., "Nonlinear, Two-Moded, Single-Fibre Interferometric Switch," *Electron. Lett.* **24**(21), 1338 (1988).
38. Black, R. J., Hénault, A., Lacroix, S., and Cada, M., "Structural Considerations for Bi-modal Nonlinear Optical Switching," *IEEE J. Quantum Electron.* **26**(6), 1081 (1990).
39. Bilodeau, F., Hill, K. O., Johnson, D. C., and Faucher, S., "Compact, Low-Loss, Fused Biconical Taper Couplers: Overcoupled Operation and Antisymmetric Supermode Cutoff," *Opt. Lett.* **12**(8), 634 (1987).
40. Lacroix, S., Gonthier, F., Houle, C., Black, R. J., and Bures, J., "Coupleurs 2×2 par Fusion et Étirage des Fibres Unimodales: Réponses Spectrales et Biréfringence," *Onzièmes Journées Nationales d'Optiques Guidées*, Grenoble, France, Recueil des communications (LEMO, Grenoble, Fax +33-76-87-69-76), 29 (1990).
41. Black, R. J., Archambault, J. L., Gonthier, F., Lacroix, S., Ricard, D., and Bures, J., "Low-Pass Demultiplexing Fused Tapered Fiber Coupler Using Residual Core Guidance: Modeling and Realization," *Workshop on Active & Passive Fiber Components*, Monterey, Calif., *1991 Technical Digest Series* **8:** Integrated Photonics Research (Optical Society of America, Washington, D.C.), 153 (1991).
42. Mortimore, D. B., "Polarization Dependence of 4×4 Optical Fiber Couplers," *Appl. Opt.* **30**(21), 371 (1991).
43. Born, M., and Wolf, E., *Principles of Optics*, 6th ed., Pergamon, Oxford, 1980.
44. Black, R. J., Veilleux, C., Lapierre, J., and Bures, J., "Radially Anisotropic Lightguide Mode Selector," *Electron. Lett.* **21**, 954 (1985).
45. Fujii, Y., "Optical Fibers with Very Fine Layered Dielectrics," *Appl. Opt.* **25**(7), 1061 (1986).
46. Gloge, D., "Weakly Guiding Fibers," *Appl. Opt.* **10**, 2252 (1971).
47. Snyder, A. W., "Asymptotic Expressions for Eigenfunctions and Eigenvalues of a Dielectric or Optical Waveguide," *IEEE Trans. Microwave Theory Tech.* **MTT-17**, 1130 (1969).
48. Snyder, A. W., and Young, W. R., "Modes of Optical Waveguides," *J. Opt. Soc. Am.* **68**, 297 (1978).
49. Snyder, A. W., and Rühl, F., "Ultra-High Birefringence Fibers," *IEEE J. Quantum Electron.* **QE-20**(1), 80 (1984).
50. Sammut, R. A., Hussey, C. D., Love, J. D., and Snyder, A. W., "Modal Analysis of Polarization Effects in Weakly-Guiding Fibers," *IEE Proc. Pt. H* **128**, 173 (1981).
51. Love, J. D., Hussey, C. D., Sammut, R. A., and Snyder, A. W., "Polarization Corrections to Mode Propagation on Weakly Guiding Fibres," *J. Opt. Soc. Am. A,* **72**, 1583 (1982).
52. Black, R. J., Theory of Single- and Few-Mode Lightguides, Ph.D. thesis, Australian National University, Canberra, 1984.
53. Olver, P. J., *Applications of Lie Groups to Differential Equations*, Springer-Verlag, New York, 1986.
54. Gagnon, L., "Self-Similar Solutions for a Coupled System of Nonlinear Schrödinger Equations," *J. Phys. A: Math. Gen.* **25**, 2649 (1992).
55. Cotton, F. A., *Chemical Applications of Group Theory*, 2d ed., Wiley, New York, 1971.

56. Stedman, G. E., *Diagram Techniques in Group Theory*, University Press, Cambridge, 1990.
57. Lax, M., *Symmetry Principles for Molecular and Solid-State Physics*, Wiley, New York, 1974.
58. Cohen-Tannoudji, C., Diu, B., and Laloë, F., *Mécanique quantique*, Hermann, Paris, 1977.
59. Montgomery, C. G., Dicke, R. H. and Purcell, E. M., *Principles of Microwave Circuits*, McGraw-Hill, New York, 1948.
60. Auld, B. H., *Applications of Group Theory in the Study of Symmetrical Waveguide Junctions,* Tech. Rep. **157,** Stanford Univ., Microwave Lab., 1952).
61. Kahan, T., *Théorie des Groupes en Physique Classique et Quantique*, Dunod, Paris, 1971.
62. Knorr, J. B., "Electromagnetic Applications of Group Theory," Ph.D. thesis, Cornell University, Ithaca, N.Y., 1970.
63. Knorr, J. B., and McIsaac, P. R., "A Group Theoretical Investigation of the Single-Wire Helix," *IEEE Trans. Microwave Theory Tech.* **MTT-19**, 854 (1971).
64. McIsaac, P. R., "Symmetry-Induced Modal Characteristics of Uniform Waveguides. I: Summary of Results," *IEEE Trans. Microwave Theory Tech.* **MTT-19**, 421 (1975).
65. McIsaac, P. R., "Symmetry-Induced Modal Characteristics of Uniform Waveguides. II: Theory," *IEEE Trans. Microwave Theory Tech.* **MTT-19**, 429 (1975).
66. Crepeau, P. J., and McIsaac, P. R., "Consequences of Symmetry in Periodic Structures," *Proc. IEEE* **52**, 33 (1964).
67. Preiswerk, H. P., Lubanski, M., Gnepf, S., and Kneubuhl, F. K., "Group Theory and Realization of a Helical Distributed Feedback Laser," *IEEE J. Quant. Electron.* **QE-19**, 1452 (1983).
68. Preiswerk, H. P., Lubanski, M., and Kneubhl, F. K., "Group Theory and Experiments on Helical and Linear Distributed Feedback Gas Lasers," *Appl. Phys.* **B 33**, 115 (1984).
69. Tjaden, D. L. A., "First-Order Correction to 'Weak-Guidance' Approximation in Fibre Optics Theory," *Philips J. Res.* **33**, 103 (1978).
70. Kapany, N. S., and Burke, J. J., *Optical Waveguides*, Academic, New York, 1972.
71. Yamashita, E., Ojeki, S., and Atsuki, K., "Modal Analysis Method for Optical Fibers with Symmetrically Distributed Multiple Cores," *J. Lightwave Technol.* **LT-3**, 341 (1985).
72. Kishi, N., Yamashita, E., and Atsuki, K., "Modal and Coupling Field Analysis of Optical Fibers wih Circularly Distributed Multiple Cores and a Central Core," *J. Lightwave Technol.* **LT-4** (1986).
73. Kishi, N., and Yamashita, E., "A Simple Coupled-Mode Analysis Method for Multiple-Core Optical Fiber and Coupled Dielectric Waveguide Structures," *IEEE Trans. Microwave Theory, Tech.* **MTT-36**(12), 1861 (1988).
74. Collin, R. E., *Field Theory of Guided Waves*, 2d ed., IEEE Press, New York, 1991.
75. Arfken, G., *Mathematical Methods for Physicists*, Academic, New York, 1970.
76. Kong, J. A., *Theory of Electromagnetic Waves*, Wiley-Interscience, New York, 1986.
77. Tellegan, B. D. H., "The Gyrator, a New Electric Network Element," *Phillips Res. Rept.* **381-101** (1948).

78. Jaggard, D. L., Mickelson, A. R., and Papas, C. H., "On Electromagnetic Waves in Chiral Media," *Appl. Phys.* **18**, 211 (1979).
79. Craig, D. P., and Thirunamachandran, T., "Radiation-Molecule Interactions in Chemical Physics," in P. O. Löwdin (ed.), *Advanced Quantum Chemistry.*, Academic, New York, 1983, p. 97.
80. Lakhtakia, A., Varadan, V. V., and Varadan, V. K., "Field Equations, Huygen's Principle, Integral Equations, and Theorems for Radiation and Scattering of Electromagnetic Waves in Isotropic Chiral Media," *J. Opt. Soc. Am. A* **5**(2), 175 (1988).
81. Engheta, N., and Philippe, P., "Modes in Chirowaveguides," *Opt. Lett.* **14**(11), 593 (1989).
82. Hopf, F. A., and Stegeman, G. I. A., *Applied Classical Electrodynamics.*, vol. 1., *Linear Optics*, Wiley, New York, 1985.
83. Kliger, D. S., Lewis, J. W., and Randall, C. E., *Polarized Light in Optics and Spectroscopy*, Academic, San Diego, Calif., 1990.
84. Stedman, G. E., "Space-Time Symmetries and Photon Selection Rules," *Am. J. Phys.* **51**, 750 (1983).
85. Yariv, A., and Yeh, P., *Optical Waves in Crystals*, Wiley, New York, 1984.
86. Altman, C., Schatzberg, A., and Suchy, K., "Symmetry Transformations and Reversal of Currents and Fields in Bounded (Bi)anisotropic Media," *IEEE Trans. Antennas Propag.* **AP-32**, 1204 (1984).
87. Vassallo, C., *Théorie des Guides d'Ondes Électromagnétiques*, vol. 2, Eyrolles, Paris, 1985.
88. Bertilone, D., Ankiewicz, A., and Pask, C., "Wave Propagation in a Graded-Index Taper," *Appl. Opt.* **26**(11), 2213 (1987).
89. Dändliker, R., "Rotational Effects of Polarization in Optical Fibers," in *Anisotropic and Nonlinear Optical Waveguides*, edited by C. G. Someda and G. I. Stegeman, Elsevier, Amsterdam, 1992, p. 39.
90. Schiff, L. I., *Quantum Mechanics*, 3d ed., McGraw-Hill, New York, 1968.
91. Mathews, J., and Walker, R. L., *Mathematical Methods of Physics*, 2d ed., Benjamin, Menlo Park, Calif., 1970.
92. Black., R. J., and Pask, C., "Equivalent Optical Waveguides," *J. Lightwave Technol.* **LT-2**(3), 268 (1984).
93. Black., R. J., and Ankiewicz, A., "Fiber-Optic Analogies with Mechanics," *Am. J. Phys.* **53**, 554 (1985).
94. Astwood, D. R., "Symmetry Analysis of Mode Frequency Splitting in Optical Fibres," final year B.Sc. (Hons.) project report 1991-2, University of Canterbury, Physics Dept., Christchurch, New Zealand, 1991.
95. Jackson, J. D., *Classical Electrodynamics*, 2d ed., Wiley, New York, 1975.
96. Sammut, R. A., "Mode Cutoff Frequencies in Elliptical and Large Numerical Aperture Optical Fibres," *Opt. & Quantum Electron.* **14**, 419 (1982).
97. Black, R. J., and Bourbonnais, R., "Core-Mode 'Cutoff' for Finite-Cladding Lightguides," *IEE Proc. Pt. J: Optoelectronics* **133**(6), 377 (1986).
98. Marcatili, E. A. J., "Dielectric Rectangular Waveguide and Directional Coupler for Integrated Optics," *Bell Syst. Tech. J.*, 2071 (1969).
99. Ladouceur, F., Love, J. D., and Skinner, I. M., "Single-Mode Square- and Rectangular Waveguides," *IEE Proc. Pt. J.: Optoelectronics* **138**(4), 253 (1991).
100. Bures, J., Lacroix, S., and Lapierre, J., "Analyse d'un Coupleur Bidirectionnel à Fibres Optiques Monomodes Fusionnées," *Appl. Opt.* **22**(12), 1918 (1983).

101. Li, W.-K. and Blinder, S. M., "Solution of the Schrödinger Equation for a Particle in an Equilateral Triangle," *J. Math. Phys.* **26**(11), 2784 (1985).
102. Shaw, G. B., "Degeneracy in the Particle-in-a-Box Problem," *J. Phys. A: Math. Gen.* **7**(13), 1537 (1974).
103. Snyder, A. W., and Rühl, F., "Novel Polarisation Phenomena on Anisotropic Multimoded Fibers," *Electron. Lett.* **19**(11), 401 (1983).
104. Someda, G. C., personal communication, 1989.
105. Waldron, R. A., *Theory of Guided Electromagnetic Waves*, Van Nostrand, London, 1969.
106. Hlawiczka, P., *Gyrotropic Waveguides*, Academic, London, 1981.
107. Someda, C. G., "A Generalization of the Symmetry Group for Fields in Gyrotropic Media," *Alta Frequenza* **36**(5), 470 (1967).
108. Fujii, Y., and Hussey, C. D., "Design Considerations for Circularly Form-Birefringent Fibres," *IEE Proc. Pt. J.: Optoelectronics* **133**(4), 249 (1986).
109. Castelli, R., Irrera, F., and Someda, C. G., "Circularly Birefringent Optical Fibres: New Proposals. Part I. Field Analysis," *Opt. & Quantum Electron.* **21**, 35 (1989).
110. Someda, C. G., and Brigato, A., "Circularly Birefringent Optical Fibres: New Proposals. Part II. Birefringence and Coupling Loss," *Opt. & Quantum Electron.* **23**, 713 (1991).
111. Someda, C. G., Castelli, R., and Irrera, F., "Modal Analysis and Design Criteria of Azimuthally Inhomogeneous Circularly Birefringent Fibers," *Opt. Fib. Sensors '88*, New Orleans, *1988 Tech. Digest Series* **2** (Optical Society of America), 281 (1988).
112. Stevenson, A. J., and Love, J. D., "Vector Modes of Six-Port Couplers," *Electron. Lett.* **23**, 1311 (1986).
113. Gagnon, L., and Black, R. J., "Fibres à Cœurs Multiples et Gaine Finie," *Neuvièmes Journées Nationales d'Optiques Guidées*, Lannion, France, *Résumés des communications* (CNET-IMP-88/5913), 34 (1988).
114. Black, R. J., Gagnon, L., Youngquist, R. C., and Wentworth, R. H., "Modes of Evanescent 3 × 3 Couplers and Three-Core Fibers," *Electron. Lett.* **22**, 1311 (1986).
115. Schecter, R. S., *The Variational Method in Engineering*, McGraw-Hill, New York, 1967.
116. Goell, J. E., "A Circular-Harmonic Computer Analysis of Rectangular Waveguides," *Bell Syst. Tech. J.* **48**, 2133 (1969).
117. Wijngaard, W., "Guided Normal Modes of Two Parallel Circular Dielectric Rods," *J. Opt. Soc. Am.* **63**(944–950) (1973).
118. Huang, H. S., and Chang, H. C., "Analysis of Equilateral Three-Core Fibers by Circular Harmonics Expansion Method," *J. Lightwave Technol.* **8**(6), 945 (1990).
119. Henry, C. H., and Verbeek, B. H., "Solution of the Scalar Wave Equation for Arbitrarily Shaped Dielectric Waveguides by Two-Dimensional Fourier Analysis," *J. Lightwave Technol.* **7**(2), 308 (1989).
120. Love, J. D., Stewart, W. J., Henry, W. M., Black, R. J., and Lacroix, S., "Tapered Single-Mode Fibres and Devices: Part 1. Adiabaticity Criteria," *IEE Proc. Pt. J: Optoelectronics* **138**(5), 343 (1991).
121. Gonthier, F., Lacroix, S., and Bures, J., "Supermode Analysis of Fused Couplers," *Photonics '92*, Montebello (Québec), Canada (1992).

122. Mortimore, D. B., and Arkwright, J. W., "Theory and Fabrication of Wavelength Flattened 1 × N Single-Mode Couplers," *Appl. Opt.* **29**(12), 1814 (1990).
123. Mortimore, D. B., "Theory and Fabrication of 4 × 4 Single-Mode Fused Optical Fiber Couplers," *Appl. Opt.* **29**(3), 371 (1990).
124. Lacroix, S., Gonthier, F., and Bures, J., "Fibres Unimodales Effilées," *Ann. Télécomm.* **43**(1–2), 43 (1988).
125. Sukuda, K., and Yariv, A., "Analysis of Optical Propagation in a Corrugated Dielectric Waveguide," *Opt. Comm.* **8**, 1 (1973).
126. Chen, Y., "Coupling of Periodic Modes," *J. Lightwave Technol.* **9**(7), 859 (1991).
127. Peng, S. T., Tamir, T., and Bertoni, H. L., "Theory of Periodic Dielectric Waveguides," *IEEE Trans. Microwave Theory Tech.* **MTT-23**(1), 123 (1975).
128. Yasumoto, K., and Kubo, H., "Numerical Analysis of a Cylindrical Dielectric Waveguide with a Periodically Varying Circular Cross Section," *J. Opt. Soc. Am. A* **7**(11), 2069 (1990).
129. Yablonovitch, E., and Leung, K. M., "Photonic Bandgaps," *Nature* **351**, 278 (1991).
130. Zhang, Z., and Satpathy, S., "Electromagnetic Wave Propagation in Periodic Structures: Bloch Wave Solution of Maxwell's Equations," *Phys. Rev. Lett.* **65**, 2650 (1990).
131. Moore, D. J., and Stedman, G. E., "Non-adiabatic Berry Phase for Periodic Hamiltonians," *J. Phys. A: Math. Gen.* **23**(11), 2049 (1990).
132. Berry, M. V., "The Adiabatic Phase and Pancharatnam's Phase for Polarized Light," *J. Mod. Optics* **34**(11), 1401 (1987).
133. Gagnon, L., and Winternitz, P., "Exact Solutions of the Cubic and Quintic Nonlinear Schrödinger Equation for a Cylindric Geometry," *Phys. Rev. A* **39**, 296 (1989).
134. Gagnon, L., "Exact Travelling Wave Solutions for Optical Models Based on the Nonlinear Cubic-Quintic Schrödinger Equations," *J. Opt. Soc. Am. A* **6**(9), 1477 (1989).
135. Gagnon, L., and Winternitz, P., "Lie Symmetries of a Generalized Nonlinear Schrödinger Equation. II. Exact Solutions," *J. Phys. A: Math. Gen.* **22**, 369 (1989).
136. Gagnon, L., "Some Exact Solutions for Optical Wave Propagation Including Transverse Effects," *J. Opt. Soc. Am. B* **7**(Special issue: Transverse effects in nonlinear optical waveguides), 1098 (1990).
137. Gagnon, L., and Bélanger, P. A., "Soliton Self-Frequency Shift versus Galilean-like Symmetry," *Opt. Lett.* **15**, 466 (1990).
138. Gagnon, L., and Paré, C., "Nonlinear Radiation Modes Connected to Parabolic Graded-Index Profile by the Lens Transformation," *J. Opt. Soc. Am. A* **8**, 601 (1991).
139. Cracknell, A. P., *Applied Group Theory*, Pergamon, Oxford, 1968.
140. Janssen, T., Crystallographic Groups, North-Holland/American Elsevier, Amsterdam, 1973.
141. Dragoman, D., and Dragoman, M., *Quantum-Classical Analogies*, Springer-Verlag, Berlin, 2004.
142. Joannopoulos, J. D., Johnson, S. G., Joshua, N., Winn, J. N., and Meade, R. D., *Photonic Crystals: Molding the Flow of Light*, Princeton University Press, Princeton, N.J., 2008.

143. Black, R. J., Lapierre, J., and Bures, J., "Field Evolution in Doubly-Clad Lightguides," *IEE Proc. Pt. J: Optoelectronics* **134,** 105 (1987).
144. Black, R. J., Bures, J., and Lapierre, J. "Finite-Cladding Fibers: HE_{12} and Local-Mode Coupling Evolution," *IEE Proc. Pt. J: Optoelectronics* **138**(5), 330 (1991).
145. Gallagher, D. F. G., "Photonic CAD Matures," *IEEE LEOS Newsletter* (Feb. 2009).
146. Dorf, R. C., *The Electrical Engineering Handbook,* 2d ed., CRC Press, Boca Raton, Fla., 1997, Chaps. 37 and 42.
147. Marcuvitz, N., *Waveguide Handbook,* 2d ed., McGraw-Hill, New York, 1986.
148. Kao, C. K., and Hockham, G. A., "Dielectric Fiber Surface Waveguides for Optical Frequencies," *Proc. IEE* **113,** 1151 (1966).
149. Culshaw, B., "Optical Fiber Sensor Technologies: Opportunities and—Perhaps—Pitfalls," *J. Lightwave Technol.* **22**(1), 39 (2004).
150. Kröhn, D. A., *Fiber Optic Sensors: Fundamentals and Applications,* 3d ed., ISA, Research Triangle Park, N.C., 2000.
151. Lopez-Higuera, J. M. (ed.), *Handbook of Optical Fiber Sensing Technology,* Wiley, Chichester, U.K., 2002.
152. Yin, S., Ruffin, P. B., and Yu, F. T. S. (eds.), *Fiber Optic Sensors,* 2d ed., CRC Press, Taylor & Francis, Boca Raton, Fla., 2008.
153. Fainman, Y., Lee, L., Psaltis, D., and Yang, C., *Optofluidics: Fundamentals, Devices, and Applications,* McGraw-Hill, New York, 2009.
154. Hawkins, A. R., and Schmidt, H. (eds.), *Handbook of Optofluidics,* CRC Press, Taylor & Francis, Boca Raton, Fla., 2010.
155. Ziemann, O., Krauser, J., Zamzow, P. E., and Daum, W., *POF Handbook: Optical Short Range Transmission Systems,* Springer-Verlag, Berlin, 2008.
156. Digonnet, M. J. F. (ed.), *Rare-Earth-Doped Fiber Lasers and Amplifiers,* 2d ed., Marcel Dekker, New York, 2001.
157. Kopp, V. I., and Genack, A. Z., "Chiral Fibers," Chap. 12 of Méndez, A., and Morse, T. F. (eds.), *Specialty Optical Fiber Handbook,* Academic Press/Elsevier, Amsterdam, 2007.
158. Ballato, J., Hawkins, T., Foy, P., Stolen, R., Kokuoz, B., Ellison, M., McMillen, C., Reppert, J., Rao, A. M., Daw, M., Sharma, S. R., Shori, R., Stafsudd, O., Rice, R. R., and Powers, D. R., "Silicon Optical Fiber," *Opt. Express,* **16,** 18675 (2008).
159. Reed, G. T., and Knights, A. P., *Silicon Photonics: An Introduction,* Wiley, Hoboken, N.J., 2004.
160. Pal, B. P. (ed.), *Guided Wave Optical Components and Devices,* Academic Press/Elsevier, London, 2006.
161. Novotny, L., and Hecht, B., *Principles of Nano-Optics,* Cambridge University Press, Cambridge, U.K., 2006.
162. Derickson, D., *Fiber Optic Test and Measurement,* Prentice Hall PTR, 1997.
163. Hill, K. O., and Meltz, G., "Fiber Bragg Grating Technology Fundamentals and Overview," *J. Lightwave Technol.,* **15**(8), 1263 (1997).
164. Someda, C. G., *Electromagnetic Waves,* CRC Press, Taylor & Francis, Boca Raton, Fla., 2006.
165. Saleh, B. E. A., and Teich, M. C., *Fundamentals of Photonics,* Wiley, Hoboken, N.J., 2007.

166. Bures, J., *Guided Optics: Optical Fiber and All-Fiber Component*, Wiley-VCH, Weinheim, 2009.
167. Agrawal, G. P., *Applications of Nonlinear Fiber Optics*, 2d ed., Academic Press/Elsevier, London, 2008.
168. Kao, C. K., "Optical Fiber Waveguide with Effective Refractive Index Profile," U.S. Patent No. 4,265,515, issued May 5, 1981.
169. Thyagarajan, K., and Ghatak, A., *Fiber Optic Essentials*, Wiley, Hoboken, N.J., 2007.
170. Bates, R. J., *Basic Fiberoptic Technologies*, McGraw-Hill, New York, 2001.
171. Kaminow, I. P., Li, T., and Willner, A. E., *Optical Fiber Telecommunications V A: Components and Subsystems*, Elsevier/Academic Press, 2008.
172. Calvo, M. L., and Lakshminarayanan, V. (eds.), *Optical Waveguides: From Theory to Applied Technologies*, CRC Press, Boca Raton, Fla., 2007.
173. Syms, R., and Cozens, J., *Optical Guided Waves and Devices*, McGraw-Hill, New York, 1992.
174. Sakoda, K., *Optical Properties of Photonic Crystals*, Springer-Verlag, Berlin, 2001.
175. Bjarklev, A., Broeng, J., and Bjarklev, A. S., *Photonic Crystal Fibers*, Springer-Verlag, Berlin, 2005.
176. Zolla, F., Renversez, G., Nicolet, A., Kuhlmey, B., Guenneau, S., and Felbacq, D., *Foundations of Photonic Crystal Fibres*, Imperial College Press, London, 2005.
177. Someda, C. G., and Stegeman, G. I., *Anisotropic and Nonlinear Optical Waveguides*, Elsevier, Amsterdam, 1992.
178. Galtarossa, A., and Menyuk, C. R. (eds.), *Polarization Mode Dispersion*, Springer Science + Business Media, Inc., New York, 2005.
179. Hadley, G. R., "Slanted-Wall Beam Propagation," *J. Lightwave Technol.* **25**(9), 2367 (2007).
180. Huang, W. P., and Xu, C. L., "Simulation of 3-Dimensional Optical Waveguides by a Full-Vector Beam-Propagation Method," *IEEE J. Quant. Electron.* **29**, 2639 (1993).
181. Yevick, D., and Hermansson, B., "Efficient Beam Propagation Techniques," *IEEE J. Quant. Electron.* **26**(1), 109 (1990).
182. Gonthier, F., Hénault, A., Lacroix, S., Black, R. J., and Bures, J., "Mode Coupling in Non-Uniform Fibers: Comparison Between Coupled-Mode Theory and Finite-Difference Beam Propagation Method Simulations," *J. Opt. Soc. Am. B* **8**(2), 416 (1991).
183. Okamoto, K., *Fundamentals of Optical Waveguides*, 2d ed., Academic Press/Elsevier, London, 2006.
184. Taflove, A., and Hagness, S. C., *Computational Electrodynamics: The Finite-Difference Time-Domain Method*, 2d ed., Artech House, Boston, 2005.
185. Felici, T., Black, R. J., and Gallagher, D., "Recent Advances and Results in Waveguide Shape Optimization Techniques," *Proc. LEOS'02* (Glasgow), Nov. 2002.
186. Black, R. J., "Consequences and Exploitation of Symmetry in the Evaluation of Optical Waveguide Modal Fields and Eigenvalues," *FIMMWAVE Tech Note*, Photon Design, Oxford, U.K., 2004.
187. Bai, B., and Li, L., "Reduction of Computation Time for Crossed-Grating Problems: A Group-Theoretic Approach," *J. Opt. Soc. Am. A* **21**, 1886 (2004).

188. Yeh, P., Yariv, A., and Marom, E., "Theory of Bragg Fiber," *J. Opt. Soc. Am.* **68**, 1196 (1978).
189. Archambault, J. L., Black, R. J., Lacroix, S., and Bures, J., "Loss Calculations in Antiresonant Waveguides," *J. Lightwave Technol.* **11**(3), 416 (1993).
190. Yin, D., Schmidt, H., Barber, J., and Hawkins, A. "Integrated ARROW Waveguides with Hollow Cores," *Opt. Express* **12**, 2710 (2004).
191. Duguay, M. A., Kokubun, Y., Joch, T. L., and Pfeiffer, L., "Antiresonant Reflecting Optical Waveguides in SiO_2-Si Multilayer Structures," *Appl. Phys. Lett.* **49**, 13 (1986).
192. Kao, C. K., *Optical Fiber Systems: Technology, Design, and Applications*, McGraw-Hill, New York, 1982, reprinted 1986.
193. Sakoda, K., *Optical Properties of Photonic Crystals*, Springer-Verlag, Berlin, 2001.
194. Kuang, W., Cao, J. R., Yang, T., Choi, S. J., Lee, P. T., O'Brien, J. D., and Dapkus, P. D., "Classification of Modes in Suspended-Membrane, 19-Missing-Hole Photonic-Crystal Microcavities," *J. Opt. Soc. Am. B* **22**, 1092 (2005).
195. Kuang, W., Cao, J. R., Yang, T., Choi, S. J., O'Brien, J. D., and Dapkus, P. D., "Classification of Modes in Multi-Moded Photonic Crystal Microcavities," in *Conference on Lasers and Electro-Optics/International Quantum Electronics Conference and Photonic Applications Systems Technologies*, Technical Digest (CD), Optical Society of America, 2004, paper CTuDD3.
196. Malkova, N., and Ning, C., "Electro-Optical Control of the Directional Switching Based on Degenerate State Splitting in Photonic Crystal," in *Conference on Lasers and Electro-Optics/Quantum Electronics and Laser Science and Photonic Applications Systems Technologies*, Technical Digest (CD), Optical Society of America, 2005, paper CTuL5.
197. Mao, W., Liang, G., Zou, H., Zhang, R., Wang, H., and Zeng, Z., "Design and Fabrication of Two-Dimensional Holographic Photonic Quasi Crystals with High-Order Symmetries," *J. Opt. Soc. Am. B* **23**, 2046 (2006).
198. Wheeldon, J. F., Hall, T., and Schriemer, H., "Symmetry Constraints and the Existence of Bloch Mode Vortices in Linear Photonic Crystals," *Opt. Express* **15**, 3531 (2007).
199. Knight, J. C., Birks, T. A., Russell, P. St. J., and Atkin, D. M., "All-Silica Single-Mode Optical Fibre with Photonic Crystal Cladding," *Opt. Lett.* **21**(19), 1457 (1996).
200. Monro, T. M., and Ebendorff-Heidepriem, H., "Progress in Microstructured Optical Fibers," *Annual Review of Materials Research* **36**, 467 (August 2006).
201. Black, R. J., Davies, T., Felici, T., and Gallagher, D. F. G., "Modeling Holey Fibers/Photonic Crystal Fibers with FIMMWAVE," *FIMMWAVE Tech Note*, Photon Design, Oxford, U.K., 2003-2004.
202. Ferrando, A., Zacarés, M., Fernández de Córdoba, P., Binosi, D., and Monsoriu, J., "Vortex Solitons in Photonic Crystal Fibers," *Opt. Express* **12**, 817 (2004).
203. Steel, M. J., "Reflection Symmetry and Mode Transversality in Microstructured Fibers," *Optics Express* **12**(8), 1497, 2004.
204. Hill, K. O., Fujii, Y., Johnson, D. C., and Kawasaki, B. S., "Photosensitivity in Optical Fiber Waveguides: Application to Reflection Filter Fabrication," *Appl. Phys. Lett.* **32**(10), 647 (1978).
205. Hill, K. O., Malo, B., Bilodeau, F., Johnson, D. C., and Albert, J., "Bragg Gratings Fabricated in Monomode Photosensitive Fiber by UV Exposure through a Phase Mask," *Appl. Phys. Lett.* **62**, 1035 (1993).

206. Meltz, G., Morey, W. W., and Glenn, W. H., "Formation of Bragg Gratings in Optical Fiber by a Transverse Holographic Method," *Opt. Lett.* **14**, 823 (1989).
207. Archambault, J. L., Reekie, L., and Russell, P. S. J., "100% Reflectivity Bragg Reflectors Produced in Optical Fibres by Single Excimer Laser Pulses," *Electron. Lett.* **29**, 453 (1993).
208. Russell, P. St. J., Archambault, J. L., and Reekie, L., "Fibre Gratings," *Physics World* **41** (Oct. 1993).
209. Kashyap, R., *Fiber Bragg Gratings*, 2d ed., Academic Press, London, 2009.
210. Mao, W., Liang, G., Zou, H., Zhang, R., Wang, H., and Zeng, Z., "Design and Fabrication of Two-Dimensional Holographic Photonic Quasi Crystals with High-Order Symmetries," *J. Opt. Soc. Am. B* **23**, 2046 (2006).
211. Morey, W. W., Ball, G. A., and Meltz, G., "Photoinduced Bragg Gratings in Optical Fibers," *Optics & Photonics News* **8** (Feb. 1994).
212. Othonos, A., and Kalli, K., *Fiber Bragg Gratings: Fundamentals and Applications in Telecommunications and Sensing*, Artech House, Norwood, Mass., 1999.
213. Hill, K. O., Malo, B., Vineberg, K. A., Bilodeau, F., Johnson, D. C., and Skinner, I., "Efficient Mode Conversion in Telecommunication Fibre Using Externally Written Gratings," *Electron. Lett.* **26**(16), 1270 (1990).
214. Bilodeau, F., Hill, K. O., Malo, B., Johnson, D. C., and Skinner, I. M., "Efficient Narrowband LP_{01}-LP_{02} Mode Converting Fiber: Spectral Response," *Electron. Lett.* **27**, 682 (1991).
215. Méndez, A., and Morse, T. F. (eds.), *Specialty Optical Fiber Handbook*, Academic Press/Elsevier, Amsterdam, 2007.
216. Ibanescu, M., Johnson, S. G., Soljačić, M., Joannopoulos, J. D., Fink, Y., Weisberg, O., Engeness, T. D., Jacobs, S. A., and Skorobogatiy, M., "Analysis of Mode Structure in Hollow Dielectric Waveguide Fibers," *Phys. Rev. E* **67**, 046608 (2003).
217. E. Desurvire, *Erbium-Doped Fiber Amplifiers*, Wiley, New York, 2002.
218. Bjarklev, A. O., *Optical Fiber Amplifiers: Design and System Applications*, Artech House, Boston, 1993.

INDEX

Note: Page numbers followed by *t* indicate tables and page numbers followed by *f* indicate figures.

www.ingramcontent.com/pod-product-compliance
Lightning Source LLC
Chambersburg PA
CBHW050458190326
41458CB00005B/1341
9780071622967